HIKING AMERICA'S GEOLOGY

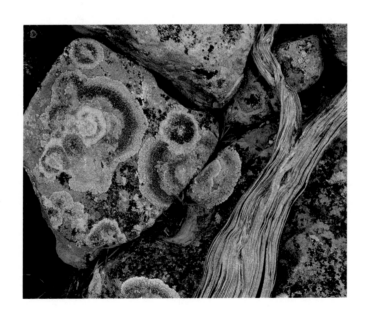

HIKING AMERICA'S GEOLOGY

by Toni Eugene and Ron Fisher

NATIONAL GEOGRAPHIC
WASHINGTON, D.C.

CONTENTS

PAGE ONE: *Lichens on rock in Desolation Canyon, Utah.*

PRECEDING PAGES: *Dawn near Mount Whitney, John Muir Wilderness, California.*

OPPOSITE: *View from the summit of Mount Desert Island, Acadia National Park, Maine.*

HAWAII VOLCANOES NATIONAL PARK: *Lava streams from a rift on the eastern side of Kīlauea, on the southeastern shore of the island of Hawaii, one of the world's most active volcanoes. Kīlauea vents have been erupting since 1983, and hikers who camp overnight in the park sometimes spot the glow of hot lava in the distance.*

ALASKA'S GLACIER BAY NATIONAL PARK: Walls of ice, crystals packed so densely that they reflect only short blue wavelengths of light, frame a hiker exploring an ice cave near the Muir Glacier in Southeast Alaska.

CALIFORNIA'S YOSEMITE NATIONAL PARK: Sunlight
silhouettes the trunks of black oaks in a meadow. Beyond them
looms El Capitan, the 3,593-foot-tall granite monolith
on the north rim of Yosemite Valley.

DINOSAUR COUNTRY: Nature paints with broad and colorful strokes in the Coyote Buttes area of Utah and Arizona's Paria–Vermilion Cliffs Wilderness, where cross-bedded sandstone—ancient sand dunes laid down in oblique layers—towers over backpackers.

MAINE'S ACADIA NATIONAL PARK: Built in 1858,
the Bass Harbor Lighthouse peers seaward from the rocky coast of
Maine. Deposits of ancient mud top granite bedrock here where
Maine's 112-square-mile Mount Desert Island meets the sea.

HAWAII'S MOUNTAINS OF FIRE

by Toni Eugene

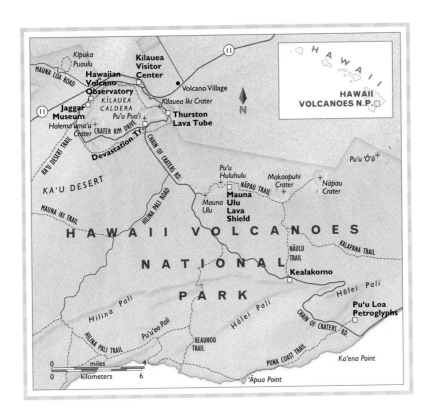

F ROM THE AIR, THE BIG ISLAND OF HAWAII IS A PATCHWORK of black, blue, and green. Black lava fields, testimony to the volcanoes that created the island, compete with bright sea, lush pastures, and tangled rain forests. Highways slice through barren moonscapes of dark rock; sheer black cliffs plummet into blue depths; sulfurous fumes and scalding steam waft from cracks in the earth. Mauna Loa and Kīlauea, two of the world's most active volcanoes, still erupt, proof that the Earth is constantly changing.

On the slopes of Kīlauea, a backpacker ascends an old lava flow of ropy pahoehoe.

PRECEDING PAGES: *Fountains of ash, lava, and toxic gases spew from Pu'u Ō'ō, a vent on Kīlauea's east rift zone, in 1986. The pyrotechnics lasted about three years and shot as high as 1,500 feet.*

Steam winds among native 'ōhi'a lehua trees. Rain forest blankets the northeastern slopes of Hawaii Volcanoes National Park, where prevailing trade winds drop some hundred inches of rain each year.

The 1,600-mile-long Hawaiian archipelago is the product of volcanoes. The islands lie in the middle of the Pacific plate, one of 16 huge slabs of rock that make up the rigid outer layer of the Earth. The Pacific plate and the other gigantic sections of rock that make up the planet's crust drift on a semiliquid layer of partially molten rock called the mantle. The island of Hawaii lies above a hot spot, an area where molten rock, or magma, melts through the Earth's crust. Heat and pressure deep within the Earth cause plumes of magma, charged with gas, to seep through the crust as lava. When eruptions occur on the ocean floor, they create fiery mountains like those that built the Hawaiian Islands.

After thousands of years and innumerable eruptions, an undersea mountain, or seamount, rises above the surface of the sea to form a new island. The stationary hot spot below the Pacific plate has been active for more than 70 million years, starting sometime in the Mesozoic and Cenozoic eras, while the plate has drifted northwest. Moving about four inches each year, it acts like a conveyor belt, carrying older volcanoes away from the hot spot, while new volcanoes build.

The Hawaiian islands are built entirely of igneous rock: rock created from magma. Kauai, formed about 5.1 million years ago, is the oldest of the main islands. In the last 750,000 years, the hot spot fueled five volcanoes, which merged to form the island of Hawaii, the chain's youngest at about 400,000 years old. Mauna Loa and Kīlauea continue the island-building process. Both produce lava composed of basalt, dark-colored igneous rock. These two mountains form the heart of Hawaii Volcanoes National Park, where volcanism is constantly on display, changing and reshaping the land. Two hundred years ago this landscape did not exist, and two centuries hence it will be no more.

Hawaii Volcanoes National Park was established in 1916, when Hawaii was still a territory. The goal was to preserve the island's volcanoes, its native plants and animals, and its cultural heritage. The park serves up wonders in myriad ways. Tour buses whisk visitors on guided tours around the park; drivers on their own can make the 11-mile Crater Rim Drive around Kīlauea Caldera or a 40-mile round-trip along Chain of Craters Road, which dead-ends, buried by lava flows. At the Kīlauea Visitor Center and also at the Jaggar Museum on the rim of Kīlauea, exhibits explain the dynamics of a volcano and display different types of lava. But the best way to see the park is on foot—a hiker can appreciate the size of the geologic formations as well as the differences between various types and forms of lava. More than 150 miles of trails in the park offer trips that range from short and easy day hikes to a several-day march straight up a mountain.

Overnight hikes require a backcountry permit from the National Park Service, so my friend Camille McNamara and I spoke to Ranger Ruth Levin at the visitor center early one Sunday in mid-May. Petite and friendly, eyes sparkling behind her glasses, Ruth advised us to sign in again when we finished the hike and to leave contact names in case something happened to us. Just a precaution: If any of the U.S. Geological Survey earthquake sensors at Hawaiian Volcano Observatory picked up vibrations that might suggest an eruption, the Park Service would send in a helicopter. Or, she suggested wryly: "If you feel a seismic rumble, just bail out." We planned to begin our backpack Monday, and she suggested several short hikes in the meantime.

Ferns, lichens, and wildflowers sprout on a fallen ʻōhiʻa lehua. The most common native trees on Hawaii, ʻōhiʻa dominate the uppermost canopy of the rain forests in the national park.

PRECEDING PAGES: Skeletons of ʻōhiʻa lehua, killed by the 1959 eruption of Kīlauea Iki, rise from among loose cinders beside Devastation Trail. Lava spatter and cinders collected downwind from the vent and formed Puʻu Puaʻi, the 400-foot-tall cone in the background.

CAMILLE AND I BEGAN TO FAMILIARIZE
ourselves with the park, starting at Crater Rim Drive, which rings the cliff-walled
summit caldera of Kīlauea, one of Hawaii's youngest and most active volcanoes.
Geologists define a caldera as a steep-walled depression at a volcano's summit. It
forms when the magma reservoir below the mountain shrinks. Unsupported, the
floor of the volcano collapses, leaving a depression. A crater, smaller than a caldera,
can form by collapse or explosion.

Kīlauea Caldera is some three miles wide and 400 feet deep; within it, Halema'uma'u Crater is 3,000 feet in diameter. We made several hiking forays from
Crater Rim Drive, all of them short and easy. Devastation Trail, about a mile
round-trip, passes over an old lava flow. White skeletons of 'ōhi'a lehua, a native
tree of the myrtle family that can grow to spectacular heights, littered black and
broken fields; bright ferns and 'ōhi'a lehua seedlings covered with pink blooms,
tiny starts of those tall trees, sprang from stark black rock. After patient searching, I found examples of Pele's tears—tiny droplets of black volcanic glass formed
as molten rock is blasted upward then cools as it falls to Earth—and fine golden
strands called Pele's hair—volcanic glass formed when thin filaments of molten
lava are carried by the wind.

Halema'uma'u, the legendary home of Pele, the Hawaiian goddess of fire, is
only a five-minute walk from Crater Rim Drive, across a field of steaming vents,
called solfataras, where reeking sulfur gases seep to the surface of the volcano.
When Mark Twain visited here in 1866, he noticed the stink of sulfur—"strong,"
he allowed, "but not unpleasant to a sinner." Sulfur crystals cling to the rock
around the solfataras, and the acidic gases color the surrounding rock. Steam escaping from one hole was so hot I had to yank my hand away. A wilted lei tribute
to Pele ringed a small boulder near the crater; a White-tailed Tropicbird soared
on wings spanning three feet above the dark gray and black expanse. Pele's home
has changed in size and depth since 1924, when it was 1,500 feet in diameter. During the last major Halema'uma'u eruption, in 1967, lava filled the crater to within
a hundred feet of the rim, then receded. A solidified circle of rock, reminiscent
of a bathtub ring, indicates how high the lava reached.

In the afternoon, Camille and I hiked the four-mile Kīlauea Iki Loop Trail.
Kīlauea Iki, or little Kīlauea, is an ancient crater that, like Halema'uma'u, col-
lapsed to form a depression when molten rock drained from below the volcano.
As we approached Kīlauea Iki, hikers below straggled like ants across seemingly
barren blackness, and wisps of steam wafted lazily upward. Steep switchbacks

Placed as a tribute to the Hawaiian fire goddess, Pele, leis brighten the rugged rim of Halemaʻumaʻu
Crater, her legendary home. During the most recent major Halemaʻumaʻu eruption, in 1967, a lava
lake filled the crater to within a hundred feet of the rim; steam and gases still seep from the floor.
Scarlet ʻōhiʻa blossoms, right, grace a lei that echoes the smooth curves of old pahoehoe.

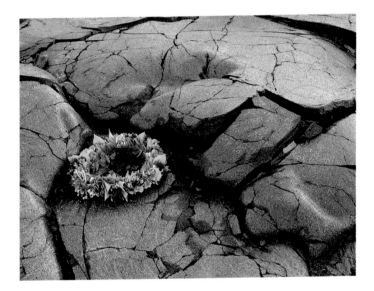

Signs throughout the park remind hikers of the dangers in this dynamic landscape. Among the first plants to return after an eruption, right, red-bloomed ʻōhiʻa seedlings and swordferns, their narrow fronds growing straight up, take root in damp cracks in pahoehoe.

led 380 feet down through a junglelike forest of tree ferns and ʻōhiʻa lehua to the crater floor. A bubbling lava lake in 1959, the floor looks now like a scene from *Star Wars* or *Planet of the Apes*: a steaming one-mile expanse of inky crust. Camille and I struggled over jumbled sections of aʻa, one of two types of Hawaii's basaltic lava. Aʻa, more viscous than the other Hawaiian lava, pahoehoe, breaks into loose, jagged pieces as it flows downhill. The clinkery rubble is sharp, and chunks roll underfoot; hiking over it demands constant attention. Pahoehoe, which contains more gas than aʻa, has a smooth and ropy surface. It brings to mind quick-frozen brownie batter, and walking on it is easier on the legs.

Two-thirds of the way across the crater loomed the gray-brown cone of Puʻu Puaʻi—Fountain Hill. In 1959, jets of escaping gas at Kīlauea Iki forced lava 1,900 feet into the air. Some lava fragments cool and solidify as they fly through the air, hitting the ground as cinders. Those that remain semimolten fall and spatter. Both cinders and red-hot spatter from the lava fountains built Puʻu Puaʻi; volcanic gases oxidized iron in the cinders, painting the cone rusty red and yellow-brown.

As we walked through clouds of sulfur smoke, Camille and I discovered that Kīlauea Iki was not at all the barren wilderness it seems from above. As lava cools

and expands, cracks form; ferns and seedlings soon take root in the crevices. The narrow fronds of swordferns and the scarlet flowers of tiny 'ohi'a lehuas were startlingly bright against their bleak surroundings. In the two hours we walked, we passed only a few hikers. Quieted by the still world around us, we spoke little. Only the wind and the sound of our boots crunching on the hard rock broke the silence.

Our warm-up hike through Kīpuka Puaulu the next day, though, was filled with sound—the drips of water falling from leaves and the songs of scores of birds calling from trees so tall they sometimes hid the sky. Kīpuka Puaulu is a bird park, and the 1.2-mile trail loops through a rain forest 2,000 years old. A kīpuka is an area of older vegetation surrounded by younger lava flows. Often a kīpuka shelters some of Hawaii's rarest native flora and fauna, because the lava that girds them serves as a barrier to invasive, non-native plants and animals. We saw some of those exotics thriving in the bird park: Wild roses hugged the trail, and beside it rose the peeling trunks of strawberry guava trees, their glossy, dark leaves widest near the tip. Yellow blossoms spiked from kāhili ginger, introduced to Hawaii some 50 years ago, and blush-pink bamboo orchids swayed on slender stalks. But narrow-leafed koa, Hawaii's largest species of tree, rose a hundred feet into the air; hard, dark-barked 'ōhi'a lehua reached heights of eighty. Feathery tree ferns, hāpu'u, flourished in the cool shade. Among all that greenery flitted birds found only in Hawaii. Honeycreepers were the two loudest. Yellowish green 'amakihi trilled as they rooted for insects and nectar. Like distant laughter, the calls of the crimson apapane—tee hee hee hee—echoed above the narrow trail.

It was starting to drizzle and noon approached, so Camille and I ate a quick snack at the historic Volcano House hotel and restaurant overlooking Kīlauea Caldera. Sepia photographs and quotations from the likes of Mark Twain adorn the walls and vie for attention with a koa wood grand piano in the lobby. I could have stayed there all afternoon as the rain pattered down outside.

But no. Camille is a determined Texan, filled with explorer's zeal. She hustled us out into the gray afternoon to start our backpack adventure, and we headed down Chain of Craters Road toward Mauna Ulu, a 400-foot-high mound, or lava shield. Magma erupts in Hawaiian volcanoes not only from summit vents but also from rift zones, areas of weakness that reach from the summit through

---

Hikers negotiate steep switchbacks to zigzag down 400 feet through lush fern and 'ōhi'a rain forest to the bottom of Kīlauea Iki Crater, then follow the four-mile loop trail as it crosses the crusted surface of a still-steaming lava lake formed in 1959.

Ferns and lichens find rootholds on a peeling 'ōhi'a lehua. Hawaiians used the dense, hard wood of these native trees for temple images and poles; exported to the mainland, it was milled into railroad ties.

the mountain flank. Since 1955, most of Kīlauea's eruptions have occurred in its east rift zone. Molten rock that piled up around one of its main vents, or areas where magma reaches the surface, built Mauna Ulu during a 1969-1974 eruption.

We left the car in the Mauna Ulu parking lot, hoisted our packs, and, in a growing drizzle, started eastward on our seven-mile-long hike toward Nāpau Crater. The first mile of the trail was easy walking, crossing pahoehoe laid down in 1973. I halted to study what looked like a stone column. From the trail guide Ruth Levin had given us, I knew it was a lava tree: When flowing lava hardens around the trunk of a living tree, it remains after the tree dies, forming a hollow pillar.

The pahoehoe grew wet and slippery in the rain. Camille and I stepped carefully. We admired the gray mass of Pu'u Huluhulu, a 400-year-old cinder cone, in the distance. Pahoehoe gave way to sharp and crumbly a'a, and the going got tougher. Relieved, we reached more pahoehoe. Regardless of the type of lava, this was a world of black and gray—dark rock stretching as far as we could see. Only cairns, called *ahu,* marked the narrow trail. These black piles of rocks in a black kingdom of more rock were hard to follow, especially as rain increased and visibility decreased. The pahoehoe got slicker as it got wetter; I slipped and swore as I tore my pants. We struggled upward, downward, and over, one foot in front of the other, alone in the world. Ponchos and hats were no defense against rain that fell harder and harder. After all, it is a rain forest. Some hundred inches of precipitation fall each year. Too bad it was all arriving in one day.

About four miles into the hike we began skirting Makaopuhi Crater, a giant double-lobed pit some 500 years old. The rain was falling so hard it was difficult to see in front of me. Our ranger guru Ruth had advised us to take plenty of water on this excursion, as none was available. We carried six quarts—12 heavy pounds!—each. With rain falling in torrents all around us, I had little desire to drink, but drink we did, partly just to lighten our load and partly to rest a while. Ahead the trees grew nearer and the terrain appeared to change.

Beyond Makaopuhi, the path wound through a dense rain forest of 'ōhi'a lehua and gigantic tree ferns that slowed but did not stop the deluge from reaching us. Mud replaced rock, and we slid into the late afternoon. Big raindrops plopped on my glasses, and we jumped wide puddles. Leaves brushed us, depositing yet more water on glasses and ponchos. The greenery was thick, and the trail was narrow. We fought our way past branches and brush, feeling overwhelmed by the foliage—yet this rain forest is young. In 1840, a pahoehoe flow flooded the entire area. After that, it looked much like the bleak terrain Camille and I had just traversed. Those tough little ferns and 'ōhi'a lehu seedlings we had admired in Kīlauea Iki make fast headway, given ample rain.

As we rounded a bend in late afternoon, we passed a rickety building surrounded by a lava-rock wall. A wooden sign noted that it had once been a *pulu* processing mill. Pulu, the golden fibers that grow on hāpu'u tree ferns, was gathered, dried, and baled here in the 1800s, then shipped throughout the Pacific and to North America for use as pillow and mattress stuffing.

The light was dimming and the rain beat down as Camille and I trudged past the remains of the pulu mill. The forest thinned, and then the trail entered a broad meadow. By dusk, the rain had slowed down to a drizzle and we stopped to adjust our packs.

"My God," breathed Camille, and I followed her gaze. To the northeast, beyond Nāpau Crater, rose a column of bright orange lava several stories high. Gray clouds,

The Nene—Hawaii's state bird, and the only survivor among several species of geese endemic to the islands—was pushed to near extinction by hunting, habitat loss, and introduced predators.

PRECEDING PAGES: *Incandescent clouds of steam and gases billow into the night as glowing lava flows tumble into the sea at Hawaii Volcanoes National Park. The current Kīlauea eruption began in 1983; it has proven to be the longest-lived rift zone eruption in recorded history.*

shot with incandescent pink, billowed around it. Sparks of red and orange and chunks of black rock fountained against a slate sky. Struck silent, we watched for several minutes. Finally, as the rain increased to a torrent and twilight approached, we hurried upslope to make camp.

It was still pouring in the near-dusk light. Everything was soggy: sleeping bags, clothes, food, us. Discouraged, we rushed to get the tent up and tossed our packs inside just as darkness fell. At the ledge facing Nāpau we checked on our volcano. The fiery column was gone, replaced by a ribbon of orange that divided the night. Even from this distance, we could see the river of lava move and change shape as it rolled down an invisible hillside. We stared, mesmerized, in the cold rain, as the light show continued. We felt no tremors. Finally, exhausted, shivering, and wet, we took shelter in our little tent.

RAIN FELL ALL NIGHT, POUNDING ON THE thin nylon. Although we were tempted to crawl out and check our volcano again, we chose semidry warmth over drenched splendor. About nine in the morning, we broke camp in thin sunshine and returned to the hike. Beyond the simple campsite area, the trail splits. We followed the branch that descends steeply into Nāpau Crater. A'a made walking difficult. Twisted lava trees, gaping cracks in the crust, and fumeroles—holes spewing hot gases—were visible. A sign at a wide bed of cinders prohibited further hiking. Beyond it loomed the gray mound of Pu'u O'ō, an 835-foot-high cinder-and-spatter cone. In January 1983 molten rock and sulfurous gases spurted to the surface at a vent just northeast of Nāpau. Over the next three years, fallout from lava fountains piled up around the vent and built Pu'u O'ō, making it the most active vent in Kīlauea's most active rift zone.

Camille is spontaneous as well as energetic. As we sloshed through puddles and clawed dripping tree ferns out of our path, she suggested that we take a spur trail, the Nāulu, to the Chain of Craters Road: "Why repeat ourselves?" she wheedled.

Opposite the twin pits of Makaopuhi Crater, we turned south on the three-mile-long Nāulu Trail. Thick rain forest bracketed it, too, and we ducked under low branches and bounded across big puddles. The sun grew stonger. The air was fresh and clean. Huge fiddleheads curled from the bases of tree ferns, and the trail sloped gently downhill. Shafts of sunlight filtered onto the path ahead. Steam rose from the trees and our bodies as the temperature rose. The rain forest ended, and again we were climbing up lava flows through a black world of cracks and crevasses. We stopped frequently to rest and drink, and I noticed again the designs in the lava. Pahoehoe had cooled in braids, loops, and sinuous curves. Chunks of a'a from the Mauna Ulu flow dotted the path and slowed us intermittently. Dusty bits of pumice, a lightweight volcanic glass full of cavities, crumbled underfoot.

The day turned hot. This hike would be brutal in bright sun. I was glad Ranger Ruth Levin had insisted we bring all that water. Again only black ahu, almost impossible to see against miles of rocks, marked the path through the wilderness.

Park rangers help interpret the wonders of the park. On a guided hike into Kīluaea Iki,
visitors learn how plants take root on new lava flows.

Tenacious ferns and 'ōhi'a lehua struggled upward through cracks large and small. In several spots the trail merged with a short section of blacktop, a reminder that when lava closed Chain of Craters Road in 1974, it had to be relocated.

At about one o'clock, Camille and I struggled out of the bushes near the Kealako-mo parking lot. A huge white and yellow tour bus was just turning around. Camille, ever Texas-expansive, waved, yelled "Hey," and charged across the road toward it. The stocky bus driver, Kalau, greeted us with a wide smile, taking in our filthy clothes and grimy faces. He was waiting to pick up a group of students who were hiking and, in the interim, would drive us to our car some six miles north.

When we checked back in at the visitor center, Ranger Kupono McDaniel con-firmed volcanic activity the night before. A flow had started on Sunday, May 12, at about 7:30 a.m., and had been named the Mother's Day Flow. Instruments picked up tremors and earthquakes on Kīlauea and near Pu'u O'ō Crater, but readings rapidly deflated. Lava flows from the southeastern base of Pu'u O'ō formed two rivers, each about ten feet wide, that were flowing rapidly downhill.

I met Ranger Kupono again the next afternoon, leading the Wild Lava Tube Tour. A lava tube forms when the top of a flow crusts over, enclosing a molten river of lava in a tunnel of hardened rock. The crust insulates the interior, and lava keeps flowing, often miles from the vent. When the eruption ceases, the lava drains away, leaving a cavelike tunnel of lava. Lava tubes are common in Hawaii, most in the park are closed to the public to protect their delicate ecosystems.

An easy 20-minute hike, though, takes visitors through Thurston Lava Tube, off Crater Rim Drive. The trail leads through a small, jungle-filled crater into the cave left by an ancient flow. Thurston is smooth-walled rock: So many people have traipsed through it that little remains of its original lava stalactites and vegetation. Once a week, by reservation only, though, Kupono leads 12 lucky tourists on a walk through another lava tube, this one discovered in 2000.

To start, Kupono handed out hard hats with headlamps. He emphasized the tube's fragility and asked that we refrain from touching anything. In the two-mile hike to the tube, Kupono pointed out 80-foot tree ferns, *māmaki*, *kāwaʻu* (a holly), and ʻōhiʻa lehua. Hawaiian plants and animals evolved in isolation, with no natural enemies, he explained. Endemic species lack the defense mechanisms of many imported organisms and quickly fall victim to them. This site is fenced, he explained, to keep out the feral pigs and goats that have destroyed much of Hawaii's native flora.

Through a skylight—a break in the roof of the tube—we climbed down a 15-foot aluminum ladder. ʻŌhiʻa lehua roots hung from the roof; rings inside indicated the levels lava had reached before receding over time. It was hard to judge distances, and I shuffled like an old crone over the rough and uneven floor. Lava walls glistened in shades of gray and silver. Delicate spiders dangled in tiny webs. Water dripped from the ceiling: Since lava is porous, rainfall percolates through it into the tubes. In dry areas, Kupono told us, Hawaiians once gathered drinking water from tubes. This tube, called Flower Head, was filled with life, evolving and changing, in contrast to Thurston's scraped-clean skeleton. Too soon, we scrambled up a rock slide to return to the sunlit day.

Viewing the vast black expanses surrounding Kīlauea now, it is hard to believe that thousands of people once lived on the shoulders of the volcano. The area was occupied into the late 1880s, when Hawaiians were gathering fresh water from lava tubes. Cultural traditions of the Hawaiian people are evident throughout the park. An easy mile walk off Chain of Craters Road leads to a boardwalked trail around Puʻu Loa, one of Hawaii's largest collections of petroglyphs. Some 15,000 carvings have been pecked into the pahoehoe here. One tradition suggests that

Caves like Thurston Lava Tube form when the lava's surface cools and crusts over
but the molten core continues to flow and eventually drains away. Tricky lighting strategies create
eerie photographs. Ancient images, right, were carved in smooth pahoehoe. As many as
15,000 petroglyphs cover Pu'u Loa and the surrounding area.

early Hawaiians came here to place the umbilical cords of their newborn babies in the rock, thus assuring them a long life. Most of the petroglyphs are cup-shaped depressions enclosed in a circle, while others are human and animal figures amid groups of holes.

Geologist Don Swanson, director of the U.S. Geological Survey's Hawaiian Volcano Observatory, guided me to another spot where lava links present to past. Capped with snow (the same precipitation that Camille and I had struggled through all day Monday), Mauna Loa loomed to the west. The sky was a bright blue, with cottonball clouds, as we started across crumbly aʻa toward the Kaʻu Desert on the leeward side of Kīlauea. The torrential rains of the east generally don't fall here in the mountain's shadow. The windswept Kaʻu is layer on layer of lava flows covered with fine brown and tan ash and punctuated with struggling shrubs.

Kīlauea exploded intermittently for 300 years before the eruptions ceased in 1790. During that time, people walked on wet ash deposits that hardened around their footprints. In 1919, when those prints were first discovered in the ashes of Kaʻu, little was known of any Hawaiian occupation of this high country, and the prints were thought to have been those of warriors crossing the desert. In recent years, however, park scientists have found abundant evidence that Hawaiians were common in the area 250 to 400 years ago. Don stopped to show me an old quarry—a flat-topped, layered rock—where Hawaiians had mined material for scrapers and other tools. Near it he pointed out three sets of footprints in the brown ash—two large and one small. "A mom, a dad, and their child, perhaps," he speculated.

We walked carefully, scanning the ground to avoid disturbing any prints. Time and again Don halted to show me a quarry site and footprints pressed into the fine-grained layers of ash near it. The ash containing prints seemed browner, harder, and a little higher than the layer around it. Gritlike gravel over the ash, he told me, helps preserve the tracks, but they are still very fragile and will erode away with time. Halfway through our hike I pointed eagerly to tracks ahead of us. They were not recorded among the several thousand sets of prints archaeologists have already mapped in the park. Don took precise notes and photographs, then recorded their exact location. He was ecstatic: "Geology, culture, and archaeology all come together in Volcanoes," he said of the park.

Don has worked at the observatory off and on for 20 years. Founded in 1912, it has monitored volcanic activity on the island since before the park was established. Its instruments detect earthquakes that might signal eruptions, trace underground magma, and study eruptions as they occur, adding not only to the knowledge of Hawaiian volcanoes but also to worldwide volcanology. Every morn-

Fiery falls from Kīlauea add new land to Hawaii's actively growing coastline. Lava and rocks ricochet skyward, and seawater boils and explodes into steam. Since 1955, most of the volcano's eruptions have occurred on its east rift zone.

ing Don stops at the ranger station where flows have buried Chain of Craters Road and visually checks lava conditions in the predawn darkness. Volcanic fumes are choking, recent lava flows collapse without warning, and lava entering the ocean creates hazardous steam and explosions. Conditions change rapidly, and rangers here and at the visitor center offer up-to-the-minute advice.

I stood on the thick pahoehoe as darkness fell and watched the distant pink glow that marked the lava's advance. Determined hikers were returning from an eight-mile round-trip over old, broken flows to see the moving pillows of lava. Their flashlights bobbed like fireflies. Like me, they had witnessed first-hand the dynamic nature of Hawaii's geology. Since the beginning of the current Kīlauea eruption in 1983, lava has wrecked more than 180 homes, buried some nine miles of highway, and added more than 500 acres of new land. The pink glow was a vivid reminder: The volcanoes that form the heart of Hawaii create as they destroy, constantly changing and reshaping the island. ▪

BLACK SAND BEACHES, PUNALUU: *Coconut palms spire above their reflections and bracket an outrigger canoe at Punaluu, on the Big Island's southeast coast; waves shine the shore. Basaltic lava that pours into the ocean chills so suddenly that it shatters into glassy particles of black sand.*

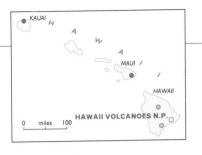

FROM DESERTS TO RAIN FORESTS, FROM BEACH strolls to rocky scrambles, Hawaii's tropical climate affords good hiking all year round, but be prepared for rain on the windward side of the mountains and hit the high spots early or late to avoid crowds.

◆ PUNALUU: Currents are strong at Punaluu, one of the most picturesque black beaches in a state famous for them. Here, hikers may see endangered green sea turtles.

◆ NA PALI COAST: Accessible only by air, by boat, or on foot, the Na Pali coast of Kauai, Hawaii's oldest island, remains a rugged wilderness of precipitous cliffs and eroded valleys.

◆ MAUNA KEA: Hikers must register for the eight-hour trek over cinder cones and lava flows from Mauna Kea State Park to the observatory at the summit of the "white mountain."

◆ HALEAKALĀ: Trails lace the erosion-gouged depression on the summit of Maui's volcano. Visitors view plants that grow only here and may also walk in to coastal waterfalls.

NA PALI COAST, KAUAI: *Barely a foot wide in places, the 11-mile-long Kalalau Trail winds up and down the sheer, sculptured lava cliffs of Kauai's northwestern coast. Many consider the path Hawaii's ultimate hike.*

MAUNA KEA, HAWAII: *Snow streaks the barren flanks of 13,796-foot Mauna Kea, where observatories staffed by people from 11 nations probe the far reaches of the universe. The dormant volcano, highest island mountain in the world, affords astronomers an average of 300 clear and cloudless days and nights a year.*

HALEAKALĀ, MAUI: *Like craters in a movie moonscape, cinder cones dot the floor of seven-mile-long Haleakalā, the ancient volcano that forms Maui's eastern rampart, near Sliding Sands Trail.*

Hawaii's Mountains of Fire

CHAPTER TWO

ALASKA'S RIVERS OF ICE

by Toni Eugene

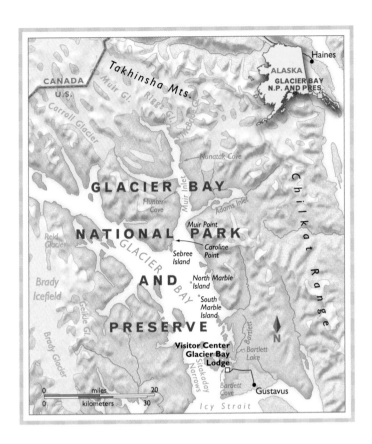

AT TWO IN THE MORNING IN LATE JUNE, IT WAS NOT completely dark. Outside the tent a ribbon of pearly white banded the eastern horizon. Our kayak was silhouetted against a sky of indigo, and the waters of Muir Inlet glinted almost black. As I leaned into a gusting wind that clawed at our tent, the sky grew slowly brighter. At about four o'clock a pale sun rose out of the dark water, and the band of light widened. I leaned harder against the billowing nylon, trying to keep the tent from blowing over. Near

Glacial striations score a rock on the shore near Reid Glacier in Glacier Bay National Park, Alaska.

PRECEDING PAGES: *A shaft of sunlight illuminates beached bergs of glacial ice in Muir Inlet. Warmer saltwater and tides melt the ice crystals; the porous remnants look white.*

Alaska's Rivers of Ice

my feet my 17-year-old son, Ted, wriggled in his sleeping bag like a Day-Glo cater-pillar. Our choice of a campsite—on the exposed edge of a rocky point—was a poor one, and we had been paying the price all night. His three-hour stint as tent stake had ended at two and mine had begun. As five raucous shorebirds began breakfasting in shallows by the beached kayak, it hit me—the water was approach-ing the kayak. "Ted," I hissed, "high tide."

"Ma," he shrieked at me as he hurled himself out of the tent. "Move!"

We had come to southeastern Alaska to explore its ice-carved fjords and shores and to see the rivers of ice that still shape Glacier Bay National Park and Preserve. At the weathered building that houses the Visitor Information Station, I had talked with Ranger Judith Challower-Wood about hiking in the park. There are no back-country trails, she told me, but beaches and areas recently released by receding glaciers offer good hiking, as do high meadows beyond the alder thickets. In the bay's west arm, people hike to Geikie Glacier and make short forays onto Reid. In Muir Inlet, some kayakers beach their boats and make the hour-long trek at low tide to view McBride Glacier; others hike inland from the beach near Caro-line Point. But kayaking, she told us, is the best way to "hike" Glacier Bay.

GLACIER BAY IS A WORLD OF WATER, which is why Ted and I found ourselves camped on a windblown beach. Like all glaciers, those of the bay formed over time as the weight of billions of snowflakes pressed into thick rivers of ice. Ice first accumulated in this part of the world about 80,000 years ago, and ice sheets have ebbed and flowed over the area ever since. During the Wisconsin Age, about 20,000 years ago, an ice sheet covered all but the highest peaks and a few headlands. About 10,000 years later the climate warmed. Forests grew as the ice retreated, only to be inundated when the ice again advanced about 3,500 years ago. Ice scoured rock downslope from the mountains; as the glaciers ground downhill, they plucked more rock, rounding the hills and lowering the valleys. Those steep-sided valleys are now the high-walled fjords of Glacier Bay. The sharp peaks that loom more than 4,000 feet above the water—atop the Takhinsha Mountains and the Fairweather and Chilkat Ranges—escaped the glacial sanding. Ice 4,000 feet thick and up to 20 miles wide extended south more than a hundred miles from the peaks of the coastal mountains into Icy Strait—until the glaciers began to retreat, around 1750.

Glacier Bay is the ancestral home of the Tlingit Indians, who were driven from the area by advancing glaciers about 800 years ago. When British explorer Capt.

Eye level with icebergs, kayakers exploring Muir Inlet keep a careful distance;
90 percent of the ice lies underwater and can puncture fiberglass kayaks.

George Vancouver sailed by in 1794, Glacier Bay was just a dent in the Icy Strait shoreline. By 1879 when John Muir, one of the first to study the glaciers here, came north from Yosemite, the ice in the bay had retreated more than 30 miles. An 1899 earthquake closed the area temporarily, but by then steamers had already carried some 25,000 to marvel at the glaciers. Glacier Bay was proclaimed a national monument in 1925, and in 1980 a 3.27-million-acre area was redesignated Glacier Bay National Park and Preserve. In 1992, Glacier Bay, Wrangell–St. Elias National Park, and Canada's Kluane National Park became part of a World Heritage site, the largest internationally protected area on Earth.

Massive living glaciers have made the park famous. The rivers of ice that lure almost 400,000 visitors a year to Glacier Bay National Park are tidewater glaciers—glaciers that flow into the sea. Huge chunks of ice break off, or calve, and tumble into Glacier Bay, Muir Inlet, and the Gulf of Alaska. While Ice Age glaciers overflowed continents, these glaciers are smaller, even the ice fields, and originate in the mountains.

Alaska's Rivers of Ice

*Brown bears romp in gilded grasses. Brown, or grizzly, bears
frequent the northern parts of Glacier; black bears inhabit the forests of the lower bay.
The park issues bear-resistant canisters to campers for food storage.*

PRECEDING PAGES: *At twilight the rugged profile of the Fairweather Range
echoes jagged spikes of fireweed blooming at Bartlett Cove, site of the Glacier Bay
Visitor Center and park headquarters.*

The park, 65 miles northwest of Alaska's capital of Juneau, can be reached only
by boat or plane. Many visitors arrive on huge cruise ships; some motor or sail in
on their own boats; others come by ferry, commercial jet, or small plane. Ted and
I flew from Juneau in a twin-engine Piper Warrior. The 20-minute flight passed low
over thickly forested hills and wide expanses of dark water. We landed in the town
of Gustavus, year-round home to 400 people, and a taxi carried us ten miles west
on the one road to Glacier Bay Lodge, the only accommodation in the park proper.

It was raining the evening we arrived—fitting, because the area is a temperate
rain forest and receives an average 75 inches of rain a year. Early the next morning, mist hid the peaks of the Fairweather Range across the bay. By ten it had

burned off, and the temperature was in the 60s as Ted and I set off to explore one of three maintained hiking trails in the park. The four-mile round-trip Bartlett River Trail stair-steps across tree roots of spruce and hemlock to the river; the eight-mile round-trip Bartlett Lake Trail branches from it and winds through spruce and hemlock forest toward its namesake lake.

Ted and I set off on the one-mile-long Forest Trail, a more tended loop just off the lodge parking lot. Spruce and hemlock soared above us. The forest of Bartlett Cove is the most mature in the bay; glaciers receded here earliest, some 200 years ago. Pale green lichens and chartreuse moss blanketed rocks and fallen trees. Moss covered nearby hummocks—rounded knolls left by retreating glaciers, composed of outwash, gravel, and sand carried by running water from melted ice. Huckleberry, blackberry, and devil's club four feet tall covered the ground. A woodpecker's thumping first broke the silence, followed by a raven's shrill screech.

To our right beyond a bend, a mallard and her five ducklings paddled across the still, dark surface of Blackwater Pond. Periodically she quacked imperiously, and her brood paddled furiously to gather for review. Blackwater is a kettle, a steep-sided depression without surface drainage that originated where a block of glacial ice rested and melted. Here the final stage of plant succession in southeastern Alaska is evident: Sphagnum moss is spreading and creating boggy muskeg. Someday Blackwater Pond will be Blackwater Bog. Steps led down past young spruce trees to the beach, where a small cruise ship lay anchored.

The visitor office, which squats by the trail below the lodge and near the dock, dispenses information, good will, wilderness permits, and PVC cylinders to keep campers' supplies safe from brown and black bears. The last two items are mandatory for anyone who wants to hike or kayak overnight in Glacier Bay, and to obtain them, visitors watch a 20-minute orientation video. Ted and I viewed the movie as a rufous hummingbird darted to and fro outside. The birds migrate all the way from Mexico to summer in Alaska.

Maya Seraphin, supervisor of the wilderness office, helped us enter the information necessary to obtain a backcountry permit so that we could leave the next morning for a three-day kayak trip up Muir Inlet, the east arm of Glacier Bay.

"What color is your tent?" she asked.

"In case they have to look for us by air, Ma," Ted explained.

I was beginning to wonder if I really wanted such an in-depth wilderness experience.

There was little time to fret, however, because we were meeting Katie White of Glacier Bay Sea Kayaks for a paddling lesson. Katie echoed the the orientation video: Tides rather than time rule any trip in Glacier Bay. Kayaking with the tide

is a breeze; bucking it is tiresome and unrewarding. The tides shift four times a day and can vary up to 25 feet, one of the highest differentials in the world. The glacier-gouged arms of the bay are narrow and emphasize changes in water level. Cross open water in slack times, Katie advised—half an hour before or after high or low tide, when the water level is static. During the two hours when the tide is actively rising or falling, 50 percent or more of the water can be rushing at once: Avoid open water then. Katie gave us a tide chart, advised us to consult it often, then helped us stow the heavy 18-foot fiberglass kayak on board the *Spirit of Adventure*, a boat that takes visitors up the bay. The tidewater glaciers are not easily accessible; the nearest are some 50 miles from Bartlett Cove in the upper reaches of the east and west arms. The *Spirit* takes tourists to view the glaciers and drops off hikers and kayakers. It makes a morning stop for those who want to explore the east arm—Muir Inlet—and cruises the west arm in the afternoon.

At six the next morning Ted and I were ready to load our gear—all of it carefully wrapped in waterproof stuff sacks or trash bags—onto the *Spirit of Adventure*. Several other kayakers were traveling the same day, and kayaks lined the bow before we finally took off with a load of day-trippers at 7:30 a.m. About 45 minutes later, in the Sitakaday Narrows, a pod of four orcas bobbed and flashed their flukes as the boat passed. One thrust its black-and-white body almost entirely out of the water to look around—an activity known as spy hopping. Nesting seagulls on one island looked like cottonballs; as the boat neared, they rose in a wave, shrieking as they circled above us. Scores of Steller sea lions draped gray rocks like limp sausages on South Marble Island.

Ted and I had decided to kayak Muir Inlet, which is narrower than the west arm of Glacier Bay. We would see no giant cruise ships here: They are banned, as are most motorized vessels. We reached the drop-off point at Sebree Island around 9:30 a.m. to find several kayakers waiting for pick-up. In minutes, our kayak and supplies, along with those of a solo kayaker from Canada, were unloaded, and those of the waiting passengers were safely stowed. It took Ted and me much longer to pack our gear and supplies into the kayak, clamber in ourselves, and— the hardest part for me—reach behind and attach our waterproof skirts.

Low tide had just passed, and water would be flowing into Muir Inlet, helping carry us north. The day was gray and overcast, and the clouds spat raindrops. We zigzagged our way up the inlet. Gray mountains draped with snow loomed beyond the rocky shore to the east. Loons, gulls, and an arctic tern or two crossed

An iceberg, calved from a tidewater glacier, competes in size with a paddler's sea kayak.

Giant chunks of glacial ice clog shallows and shore near McBride Glacier.
Kayakers can enjoy the music an iceberg makes as it melts: the drip of water,
the pop of air escaping, and the crack as it breaks apart.

PRECEDING PAGES: *Steller sea lions, loud and gregarious, haul themselves onto rocks throughout the*
bay. The north side of South Marble Island remains the best place to see them congregate.

the bow. Two harbor porpoises poked their gray heads out of the green water and gazed serenely as we paddled past. Literally on top of the water and humbled by the scale around us, we gazed up at massive headlands on the western shore, our kayak lost in the immensity all around us.

The geology of Glacier Bay is complex. In this part of Southeast Alaska, two of the plates that make up Earth's crust jostle against each other. Over the ages, bits of the Pacific plate have broken off and melded onto the North American plate. These bits are called terranes: masses of a particular kind of rock bounded by faults. An estimated five terranes combine to form this part of Alaska. An area of 400-million-year-old limestone from a coral reef may abut an entirely different sediment, 200 million years younger. The headlands we passed were part of the Alexander terrane, which geologists think may have formed in a tropical sea.

We stopped to adjust the rudder about noon, and Ted and I switched places. Now we seemed to fly over the water. A sea otter popped its grizzled head out and peered at us from shoe-button eyes, so close we could see its nose quiver. A harbor seal—its black eyes huge in its little round head—regarded us gravely.

Ahead on the eastern shore, a dirty white river of ice striped with black—McBride Glacier—slipped in and out of view as we paddled north. Around four o'clock, as high tide arrived, the wind picked up and the going got harder. We pulled into Hunter Cove. Immediately, an army of mosquitoes and gnats descended. Maxi Deet proved no deterrent; each of us was surrounded by a thick, black cloud. As we unloaded the kayak and carried our gear over the rocky beach, it started to rain. It poured as we hauled the boat above the high-tide line and tied it to an alder. It poured as we struggled to pound tent pegs into the rocky shore. And it poured as we heated and ate our dinner. Wet, tired, and discouraged, we ate quickly and took refuge in the tent.

The next morning the snow-streaked gray crags on the east side of Muir Inlet rose above banks of white clouds. Gauzy mist hugged the shore. At least it wasn't raining. Exploring beyond our campsite, I found thick brush. Sitka alders ten feet tall grew together; the branches interlocked, forming a barrier reminiscent of the palisade of a medieval castle. I clawed my way into the thicket, where spiny devil's club stabbed me and branches tangled in my hair. No hiking here without gloves and a machete or an ax. I gave up the fight and returned to the beach.

As we paddled toward McBride that morning, we saw our fellow kayaker from the drop-off boat, the first human we had encountered. At about 1:30, waves began forming and quickly grew into a heavy chop. I checked our tide chart. The cycle was about halfway between high and low tide, the time Katie had warned us was a period of rushing water. The kayak bounced over growing whitecaps, and Ted and I dug in our paddles to reach calmer waters in Nunatak Cove. We sped past stumps of an interglacial forest. About 3,500 years ago, advancing glaciers engulfed forests of spruce and hemlock, which then covered much of the hills and valleys. When the glaciers began to recede 200 years ago, outwash streams cut through the ice, revealing freeze-dried trunks, stumps, and roots of relics that resemble tortured chunks of driftwood.

Reaching the protection of Nunatak Cove, we discovered five more kayaks there, also on the way to McBride, waiting for calmer water. A nunatak is a mountain or hill completely surrounded by glacial ice. The round one for which this cove is named pokes abruptly nearly 1,200 feet above the flatter land all around. With no downpour to deter me, I studied the beach—itself a glacial remnant. The cobbled shore was made of till, a mix of sand, gravel, and boulders left by a glacier. In Glacier Bay, a shore level and wide enough for beaching a kayak is a moraine: earth and stones carried by a glacier and finally deposited when it retreats. Most are ground moraines, fairly flat areas of till. The till of the beaches is a foot deep; underneath it lies bedrock. Small wonder it was hard to set up our tent.

Muir Inlet was calm as we left the cove; the tide was with us, helping us north. As we neared McBride, though, the water grew choppier, and we paddled hard into a cold wind. Chunks of ice, some worn and rounded like sculptures by Henry Moore, floated past. We rounded a point to find the glacier on our immediate right.

Even at a distance of two miles, McBride Glacier dwarfed the kayak. A dark streak clearly divided the dirty white and light gray ice that made up its top. As flowing ice moves downhill, lateral moraines form along both edges. When two glaciers merge, the lateral moraines join into a single dark strip of till called a medial moraine: the dark ribbon that bisected McBride.

McBride's snout—the terminus of a glacier—glinted blue, like the hottest part of a candle flame. Glacier ice is dense and contains little air. The large crystals absorb all the colors in the visible spectrum, reflecting only the short blue wavelengths. As we looked, thunder boomed. It took a moment to realize it was white thunder, the sound of ice calving from a glacier. Ted and I inadvertently paddled over a low, two-foot-long piece of ice and cringed as it thumped heavily under the thin fiberglass between us and the 38-degree-water of the inlet. Kayakers must take care; most of the ice is underwater.

McBride now lies at the end of a mile-and-a-half-long fjord of its own making. Thirty years ago, the glacier was a solid wall of ice that emptied directly into Muir Inlet. Slack tide was approaching, so Ted and I paddled between the twin prongs of land that bracket the fjord. We maneuvered carefully through a maze of larger and more numerous icebergs. McBride towered some 150 feet above the water, and its snout stretched half a mile.

The glacier seemed alive, sighing, creaking, and groaning. Cracks like gunshots sounded. Crevasses, fissures hundreds of feet long in the face of the glacier fractured the white wall. Sunlight glittered on toothlike seracs, jagged pinnacles of ice. Frozen caves loomed large, tiny themselves in the frozen jumble that supported them, and some of the calved burgs were van-sized. Even from a quarter of a mile away—the closest any kayak should safely venture—the glacier dwarfed everything around it, including the harbor seals that rode the ice.

Ice art: Water, wind, and sunlight shape ice into modern sculptures.
Stranded at low tide on a rocky beach near McBride Glacier, an iceberg white with age and
air pockets resembles a work by Henry Moore. Large icebergs may take more than a week to melt.
Fresh glacial ice shines clear and blue.

Eerily alight, an ice cave near Muir Glacier lures a hiker. Danger abounds in grottoes
on the edge of glaciers: Without warning, slabs of ice give way and rocks drop from the ceiling.

PRECEDING PAGES: *"White thunder" booms as ice calves from the snout of McBride Glacier.*
Water undermines the ice at the front of tidewater glaciers, and masses up to 200 feet high
break away and crash into the sea, creating enormous waves.

A TIDEWATER GLACIER ADVANCES ONLY IF it has plucked enough rock debris from upslope to dump and build a protective shoal at its snout. That shoal, the glacier's terminal moraine, extends underwater beyond the surface of the glacier, providing a buffer between the ice and the seawater that melts it. McBride, in retreat, has left a gravel spit extending into a fjord rimmed with gray sand. The ice has so recently scoured the land that little has had time to grow. The plants are simple lichens and mosses. There are no real trails. On either side of the inlet, kayakers can beach their boats and scramble in on foot to look at the glacier, about an hour trip each way, but the going is rough and best done at low tide.

The day had turned bright and sunny, and the tide was flowing our way—in. Ted and I decided to paddle on to Riggs Glacier, about three miles up the inlet.

Yellowing and a bit bedraggled, Riggs reminded me of an aging movie star still trying to woo an audience. A tiny white-and-green sailboat lay at anchor on one side of the inlet, almost invisible against acres of ice. The receding ice revealed a gigantic brown piece of bedrock near Riggs's base. Kayakers can land here and walk in toward the glacier, but few are foolhardy enough to stay long. Glacier hiking requires specialized equipment and experience. Creaks, groans, and pops remind the visitor that it is dangerous to linger.

Less than 50 years ago, Riggs and Muir Glaciers, the latter named after the pioneering conservationist who would make the bay famous, were connected. Since then, both have receded, and Muir is tucked back in the inlet several miles west of Riggs.

Ted and I headed back down the inlet at about five p.m. The evening was a beautiful one—soft light, the wind at our backs, loons calling, and the tide carrying us south with it. As we passed McBride, thunder boomed again, and a fleet of icebergs in shapes like swans and anvils escorted us. A bald eagle, white head bright in the muted light, lifted off the branch of a tree and flapped low and leisurely over green water shifting to blue. We stopped for dinner—heating then wolfing down prepackaged chili—and considered setting up camp. But the evening was too pretty to waste; we decided to go on, slowing to admire the changing colors and shifting light. The sky pinkened, and the blue water took on a lavender cast. Another eagle soared above us. Flotillas of foot-long black birds called pigeon guillemots bobbed companionably nearby. As we approached Hunter Cove, Ted and I realized we had paddled more than 20 miles in one day. Stiffly, I hauled myself out of the kayak. Once again, the rocky shore defeated our attempts to seat tent pegs securely, and we braced them with big rocks from the beach. The sleeping bags seemed luxurious.

Achy and lazy, Ted and I puttered around camp the next morning, warmed by a bright sun. The sky was a vivid blue—not a cloud scarved the peaks on the opposite shore—and the temperature was in the high 60s. It was almost noon before we hauled our gear and kayak down to the water. Heading south toward the Sebree pick-up, we paddled against an incoming tide, not strenuous work but slow going. I tensed as the kayak suddenly shifted.

"Ma," whispered Ted urgently. "See that?"

Held captive by the waterproof skirt, a kayaker cannot turn around. I had not seen Ted's "that." Then I heard a loud whoosh as a huge column of air and water erupted about a hundred yards in front of us.

"An orca?" Ted asked. Too big, I thought.

We stopped dead—struck silent—as the sighing exhalation and accompanying cylinder of air and water appeared again, and once more. A few minutes later,

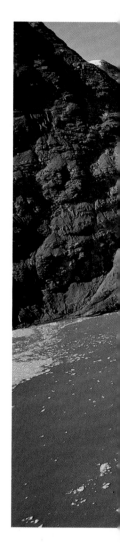

Rapidly retreating, McBride Glacier recedes about 250 feet every year.
In 1978, left, hikers gazed on a river of ice. Just five years later, right,
at the same spot, the snout of McBride had shrunk back noticeably.

Alaska's Rivers of Ice

---

the enormous black tail fin of a humpback whale flashed as it dived into the inlet. About 35 of the endangered whales summer in Glacier Bay (the number varies from year to year). We could not judge the length of this one, but most are about 50 feet long. Ted, who tended to paddle mid-channel, connecting destinations with straight lines, had no aversion to hugging the shore for a while.

As the afternoon wore on, whitecaps formed in open water and waves slapped the shore. We hauled out at about three and took turns napping while the water calmed a little, then ate dinner. The bay was still choppy but we wanted to get as close to our take-out as possible. Camping is forbidden within a mile of the drop-off point, and a couple of good sites had already been taken, which is how we found ourselves at the edge of Caroline Point. Not a great campsite, we knew, too rocky and too exposed. But it was 7 p.m.; we were tired and growing cranky. As soon as we had set up the tent and were comfortably settled, the wind picked up. By 11 p.m. it was howling and had ripped the rain fly loose, so Ted got up to lean against the nylon and hold the tent in place. Then he slept fitfully during my shift, while I admired the play of light and dark outside the tent. By five it was bright enough to study the crow-sized oystercatchers. Black with pink legs, bright orange bills, and orange rings around yellow eyes, they nattered at each other and poked in shallows getting narrower in the rising tide. That's when I woke Ted and we sprang into action. He rescued our bear canisters, which were floating away with the tide, and I struck the flapping tent. In record time—less than ten minutes—we were packed and paddling.

Sebree seemed crowded after our days in the wilderness. By 9:30, when the *Spirit of Adventure* arrived, there were 18 of us milling around and 10 kayaks waiting to be hoisted aboard. Again the exchange was effected with dispatch. Ted and I adjusted to civilization immediately—cold soda for him, hot coffee for me—and settled back to enjoy a tour up the west arm. Kittiwakes nested on sheer cliffs. A naturalist on board pointed out horizontal gouges left by glaciers on the rock walls near Rendu Inlet, then noted the black lines left by a terminal moraine.

*Soaring on wings that can span seven feet, bald eagles dive at up to a hundred miles an hour
to pluck fish from the rich waters of Glacier Bay. Common in the park, where they are
year-round residents, these eagles remain endangered in the lower 48.*

From the boat, Glacier Bay's scale had changed. The scenery was still amazing—
but diminished somehow. An eagle and its eaglets were barely visible in their distant nest in a cottonwood. A mountain goat was a white blob on a high knob.
Grand Pacific and Margerie Glaciers loomed at the end of Tarr Inlet, dwarfing a
cruise ship anchored nearby. The icy front of Margerie seemed smaller than McBride.
Missing were the sighs, rumbles, and creaks that made it seem alive. When a huge
piece of ice calved, I recognized that it would have swamped our trusty kayak; yet
the little berg we had bumped over the day before was more memorable. From
high above the water Ted and I watched the breathing spouts of two humpbacks.
They recalled our own heart-stopping encounter at close quarters, and we were
grateful we had chosen to explore Glacier Bay from the water. Kayaking gave us
an appreciation of all the park's wonders: its wildlife, its scale, its wildness, and the
ever-changing rivers of ice that carved it and constantly reshape it. ■

MISTY FIORDS: *Thick stands of hemlock, spruce, and cedar alternate
with muskeg in Southeast Alaska's Misty Fiords National Monument.
To explore this temperate rain forest wilderness, visitors hike a boardwalk trail
about a mile from Punchbowl Cove to Punchbowl Lake.
Punchbowl Creek, right, foams with runoff.*

GLACIER BAY, ALASKA, IS ONE OF NORTH
*America's wettest regions, its precipitation giving
rise to its ever-changing glaciers. The area's rugged
mountains, pristine wilderness, and rivers of ice
draw sea kayakers, hikers, and climbers alike.*

◆ MISTY FIORDS: Just a few trails thread this
2.3-million-acre national monument, accessible pri-
marily by boat or floatplane. Few facilities exist,
but visitors can reserve Forest Service cabins.

◆ ALSEK LAKE AND RIVER: Fed by snowmelt
and glaciers in British Columbia, the Alsek offers
white-water wilderness trekking through the
northwestern corner of Glacier Bay National Park.

◆ SAINT ELIAS MOUNTAINS: Crisscrossing 250
wild miles, from Canada's Yukon Territory
through Glacier Bay, these mountains spawn some
of Southeast Alaska's great tidewater glaciers.

◆ COLLEGE FIORD: Cruise ships ply the waters
of Prince William Sound. Outfitters offer kayak
rentals and guided tours of Ivy League–named
tidewater glaciers.

ALSEK LAKE: Fresh bear tracks mark a sandbar in Alsek Lake, in the northwest corner of Glacier Bay National Park. House-size icebergs dot the far shore, calved from the five-mile-wide face where the Alsek and Grand Plateau Glaciers join.

SAINT ELIAS MOUNTAINS: *Steep green slopes of Southeast Alaska's St. Elias range rise sharply beyond a field of fireweed and Indian paintbrush. Highest coastal range in the world, the trackless wilderness of these mountains challenges expedition hikers and climbers.*

Alaska's Rivers of Ice

COLLEGE FIORD: *Stranded chunks of glacial ice glow clear and blue in College Fiord, on southern Alaska's Prince William Sound. The U.S. Forest Service maintains cabins for campers who fly or boat in to this remote area.*

CHAPTER THREE

California's Yosemite Jewel

by Toni Eugene

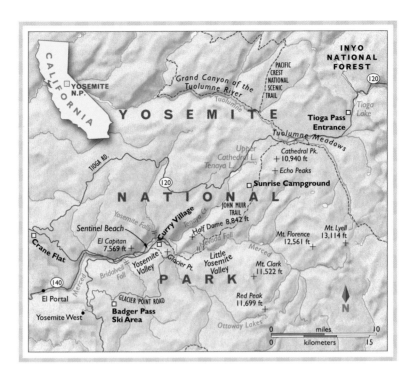

THE LEAPING WATERFALLS, MASSIVE GRANITE DOMES and spires, steep cliffs, and majestic monoliths of Yosemite Valley have lured artists, writers, geologists, and tourists since 1851, when Maj. James Savage and his militia pursued Indians into the gorge. Yosemite's scenery is so spectacular that just 13 years later, Abraham Lincoln signed a bill preserving the valley "for public use, resort, and recreation . . . for all time." Only one mile wide by seven miles long, California's Yosemite Valley showcases its geologic splendors

Soft light evokes the tranquillity of Tuolumne Meadows in Yosemite National Park.

PRECEDING PAGES: *A highland pond reflects the distinctive profile of Cathedral Peak. It juts above the trimline, the highest point scoured by glaciers 15,000 to 20,000 years ago.*

California's Yosemite Jewel

in a small and easily accessible area. But the valley, like a perfect diamond in a spectacular necklace, is so compelling that the rest of its setting is often ignored. The necklace is the Sierra Nevada, a 400-mile-long mountain range formed over millions of years as a block of Earth's crust uplifted and tilted westward. The range, made mostly of types of granite, stretches along California's eastern flank.

Yosemite National Park was set aside in 1890 to preserve a 1,200-square-mile portion of the central Sierra Nevada. Seven square miles of the park, Yosemite Valley, is a geologic gem carved and scoured by glaciers and erosion. But the Sierra high country makes up 94 percent of the park, and it is from the subalpine meadows and granite overlooks that the hiker can appreciate the grandeur and scale of the spectacular waterfalls, towering cliffs, and sculpted domes that prompted John Muir to call Yosemite the "Incomparable Valley."

Because the valley is so contained, hikes usually last only a day. More than 800 miles of trails lace the park, and to get a taste of its variety I spent a few days walking short paths and admiring the soaring cliffs, flat-floored valley, and shimmering cascades that leap into space. Then I hiked down from the meadows of the high country into the granite gorge of the valley.

T HE FACETED WONDERS OF YOSEMITE National Park's geology evolved slowly. Five hundred million years ago, during the Paleozoic era, an ocean covered the area we now call the Sierra Nevada. Over time, thick layers of sediment were folded, twisted, and eventually lifted above sea level. During the succeeding Mesozoic era, the North American plate collided with the ocean plate to the west. Results of this collison were volcanic eruptions on the surface, and bodies of molten rock—magma—were intruded into the continental crust. These bodies of magma, which are called plutons when they crystallize and solidify deep underground, now comprise the Sierra batholith. The heat from the plutons changed the country rock, the host into which they intruded, made up of seabed sediments deposited many years earlier. Erosion has now worn away nearly all the country rock, leaving the plutons exposed.

Throughout the park at least seven different types of granite can be identified. Granitic rocks are composed of three major minerals: quartz, potassium feldspar, and plagioclase. The quantity of each in a particular rock determines whether it is classified as granite, granodiorite, or tonolite. Granite, for example, is composed of almost equal parts of the three minerals, while granodiorite has more quartz and plagioclase but less potassium feldspar. It's interesting to note the different

*Larger than life, a two-inch-long dragonfly hugs a blade of grass
in a swampy area off the John Muir Trail near Little Yosemite Valley.*

California's Yosemite Jewel

California's High Sierra

---

granitic rocks while hiking along the various Yosemite trails: There is El Capitan granite, with its salt-and-pepper look, for example, and Cathedral Peak granodiorite, from the high country, with its distinctively large, light-gray crystals of potassium feldspar.

With headwaters in the high country, the Merced River twines through Yosemite Valley. Fifty million years ago, it flowed through a broad valley. Some 25 million years ago the uplift and tilt that gave the Sierra Nevada their present shape began. Three million years ago, the Merced had carved a narrow, V-shaped canyon 3,000 feet deep. Then, during the Pleistocene age, beginning about two million years ago, a series of icefields covered much of the high country and glaciers moved down many of the valleys. Boulders and rubble quarried by the moving ice scoured and sculpted the land. Half Dome jutted above the ice as glaciers gouged Yosemite Valley. Ice fractured exposed rock along joints, breaking away weaker sections but leaving harder portions, as now seen in Half Dome and El Capitan. Sheer walls, rounded domes, hanging valleys, and a U-shaped valley evolved as glaciers moved through the canyon of the Merced River.

Some 60,000 to 30,000 years ago a smaller ice sheet—glaciers of the Tioga Icefield—advanced into the valley, widening and deepening the floor, polishing rocks, and shaping the land. When the last valley glacier melted, about 10,000 years ago, the accumulated rocks and boulders it left as moraine dammed the river and created a lake. Eventually, sediment filled in that lake to build the flat floor of today's valley. Beyond the valley is a wild filigree of flower-strewn meadows, oak woods, and evergreen forests of Douglas fir, incense cedar, and ponderosa pine.

JIM SNYDER, PARK HISTORIAN, WELCOMED me into his office in the park's headquarters to talk about the area's geology. In more than 40 years with the park, starting on a trail crew, he has built an appreciation and understanding of the area that make him one of its finest interpreters. "Indians had been living in Yosemite for nearly 4,000 years before the first non-Indians entered in 1851," he told me. The last Indians in the valley were Miwok. They were gone, though, by 1855, as the first tourist party descended to marvel

June dewdrops spangle phlox in Yosemite high country's Long Meadow,
near the Sunrise High Sierra Camp. Flowers in the High Sierra begin to open in June,
but lowland plants begin to bloom in May, with some blooms lasting into autumn.

at the geologic wonders of Yosemite. Jim, pensively fingering his luxurious gray muttonchop whiskers, advised me to see the high country as well as the valley. "Cathedral Pass was one of the corridors geology created through the Sierra Nevada," he told me. "Indians used it for centuries. Hike to Glacier Point or take the trail up to Sentinel Dome. Most visitors come to the valley, but they see only part of Yosemite."

I went ahead and focused first on the popular valley, though, joining the crowds on the shuttle buses that distribute the park's visitors—more than three million a year from around the world. The day was sunny and warm, almost cloudless. The bus was packed and filled with conflicting conversations. In the seats behind me, a group of French tourists agreed that *"il fait beau."* From the bus we could see Half Dome looming at bends in the road and Yosemite Falls flashing beside the huge parking lot at Curry Village. At stop number 11, I got off with a crowd of other people intent on seeing the view of Half Dome made famous by master photographer Ansel Adams. Rafters bumped through white riffles in the green, boulder-strewn Merced. Beyond a bright meadow filled with tall white blooms of

cow parsnip, the two major cascades of Yosemite Falls sparkled and twisted as they plummeted from gray cliffs. I have admired Adams's famous photograph hundreds of times and had already had several glimpses of Half Dome. And still, I was struck silent by the view. Even with people jostling me as they passed by, the gray granite face limned with darker rock was breathtaking—huge, powerful, and moving.

Half Dome, on the eastern end of Yosemite Valley, seems ever present. It looms to an elevation of 8,842 feet, or 4,758 feet above the valley floor at Mirror Lake. Like many other domes in the park, its rocks are plutonic—and, at 87 million years old, it is composed of the youngest granitic rocks in the valley. Its missing half may have been quarried by frost. As the granite that formed this and all the park's monoliths and domes cooled under pressure, it fractured. On close inspection, fracture patterns are visible in the rocks. Glaciers originally shaped the domes, and frost, cold, heat, rain, and snow enhanced their design. All these forces have eroded the rocks into domes.

SOME STREAMS FLOWING INTO YOSEMITE Valley have persisted through periods of uplift and glaciation. They are now the spectacular waterfalls found on the valley's rim. These falls were first created when the main glacier dug deeper than the smaller, higher, tributary glaciers. For millions of years streams entered the valley from the sides and cascaded over intermediate ridges. The trunk glacier dug deeper than the tributaries did, scouring away the ledges. When the main glacier melted, it left tributary valleys hanging high above the trunk valley, leaving streams such as Yosemite and Bridalveil Creeks now falling headlong into space.

Yosemite Creek pours from a sheer cliff and falls 2,565 feet as Upper and Lower Yosemite Falls. The hike to the lower falls is just a mile round-trip, about 20 minutes. Alongside the trail, huge granitic boulders split vertically are testimony to the working of water on joints and fractures in the rock. It was a bright day, over 80 degrees, when I hiked to Yosemite Falls, but suddenly the temperature dropped

---

Scaly bark clings to a dying pine near Echo Peaks; a lightning strike may have ignited the heart of the tree. John Muir called the coniferous forests of the Sierra "the grandest and most beautiful in the world."

Fiery clouds highlight the distinctive profile of Half Dome. Joints—cracks found as sets of parallel fractures in the rock—form zones of weakness in otherwise erosion-resistant granite. Half Dome's sets of joints are vertical, which allowed the erosion of its steep northwest face.

as I approached the falls, and a cold breeze prompted me to put on my jacket. Beyond a bend I heard the roar of the falls, then turned to see water shooting several feet out into the air. Mist blew toward me, coating my face and dampening my hair. Beyond the bridge, over the creek, warm light painted gray cliff walls.

The next day I found 620-foot Bridalveil Fall just as spectacular, if on a smaller scale. The short walk from the parking lot to Bridalveil brings you closer to the actual cascade than you can get at Yosemite Falls. It is quieter and somehow more graceful, too, with a delicate silver free fall that bends in the wind.

Beyond Bridalveil I left the confines of Yosemite Valley and drove the 32 miles to Glacier Point, past the Badger Pass Ski Area. In that hour-long drive I passed by meadows dotted with purple lupines, orange poppies, and pink monkeyflowers. Before the Wawona Tunnel, I stopped at a turnout alongside several tour buses to look back and appreciate the view—Yosemite Valley, with El Capitan, Half Dome, Sentinel Rock, Cathedral Rocks, and Bridalveil Fall all spread out before me.

The view when I reached 7,214-foot Glacier Point—3,200 feet above the valley floor—was even more astounding. East beyond the valley the Sierra Nevada stretched to the horizon. I walked about a quarter of a mile from the parking lot to the geology hut, where information about Yosemite geology is provided. Golden brodiaea almost three feet tall waved in the breeze. There was so much to see from the edge of the cliff, it was hard to take it all in.

Below me I could see the swimming pool of Curry Village, one of the main lodgings in the park, and beyond it, the stone walls of the Awahnee Hotel, the park's most luxurious and historical hotel. To the east, the Merced's Nevada and Vernal Falls gleamed white and pewter as they cascaded; sunlight played across Half Dome. Thirty miles farther east clouds draped the jagged summits of Mounts Florence and Lyell. Across the valley tumbled the two silver tiers of Yosemite Falls. To the left, at the northern rim of the valley, jutted the pale granite monolith of El Capitan. It rises 3,500 feet sheer and straight from the valley floor, attracting rock climbers from around the world. Through binoculars I scanned its wrinkled face: I was too far away to see climbers, but I knew they were there, daring routes so daunting they have garnered names like "Flying Monkeys," "Lurking Fear," and "Pacemaker."

The Four Mile Trail connects Glacier Point to Sentinel Beach in Yosemite Valley. It switchbacks down 3,200 feet in 4.6 miles, past boulders and through live oaks to the trailhead on the Merced at Sentinel Beach. Lazy (and saving myself for the three-day hike to come), I chose to wander the trail downward toward the valley for an hour on a perfect June day. Leaving the crowds oohing and aahing at Glacier Point, I followed a paved walkway that soon became a narrow dirt path hugging the lip of a cliff. To the right, the land fell away to the valley shot with sunlight. I hiked through stands of pines and cedars, passing through sunshine and shadow, shimmering heat and damp coolness. Yosemite Falls sparkled again on the far side of the valley; the Awahnee crouched below gray steeps. I rested a while in breeze-swept shade beneath a red-barked ponderosa pine, then headed uphill to Glacier Point. A taste of the trail was perfect that day, and I was ready to start my high country hike in the morning.

Although Yosemite National Park contains 13 campgrounds, 4 of them in the valley, the National Park Service requires permits for all overnight camping in the backcountry. They are available 24 weeks in advance, and they go quickly. I applied for one in March for our June trip. With permit in hand, I drove west an hour

and a half—55 miles—along the Tioga Road from Crane Flat, near the northwestern entrance to the park.

The road, which usually opens in June and closes in late fall with the first major snow, winds steadily upward to Tuolumne Meadows, at 8,575 feet the largest subalpine meadow in the Sierra Nevada. Turnouts along the route offer spectacular views of the valley and arrays of granitic domes, but I resisted the temptation to stop until Tenaya Lake. It sparkled bright blue against the shining gray rock of Tenaya Peak, Pywiack Dome, and Tresidder Peak. Fine rocks and sand frozen in tons of ice scraped and polished the sweeping walls and domes beyond Tenaya. They actually shone in the bright sunlight.

OUR HIKING PARTY GATHERED AT THE ranger station and packed our supplies. Sallie Greenwood, researcher for this book, is an avid hiker and climber, as is her friend, geologist Lin Murphy, who joined us. My brother-in-law, Mac Brown, escaped his Washington, D.C., commute to help shoulder our load.

While Mac and Sallie sorted gear, Lin took me to Tioga Pass, at 9,945 feet California's highest automobile pass, near the eastern entrance to the park. She showed me examples of red and gray metamorphosed sedimentary and volcanic rocks: the country rock into which the plutons intruded. Our hike from Tuolumne Meadows to Yosemite Valley, she told me, would illustrate the earliest geology of the park and mark the route of glaciers. Lin and I rejoined Mac and Sallie a little after noon. The day was perfect—bright and sunny with a soft breeze. We shouldered our packs at Budd Creek Trailhead and stepped onto the John Muir Trail at an elevation of more than 8,000 feet.

The path rose gently upward at first, past lodgepole pines, incense cedars, and sweeping views of granite domes. Easy, I thought. I can do this. Then the angle increased, and I found myself leaning on my walking stick as we tramped three quarters of a mile—seemingly straight up. With the 600-foot increase in elevation I started to pant; perhaps this wouldn't be so easy. The trail leveled out just in time for me and entered a long emerald meadow. In the distance a mule deer

Common in higher elevations along the John Muir Trail, fleshy snow plants
are saprophytes: plants that feed on decayed matter and contain no chlorophyll.
A member of the wintergreen family, the snow plant can grow to about
a foot tall in the mountain forests of California, Oregon, and Nevada.

placidly munched the new grasses. The meadow was boggy; we skirted puddles. Few wildflowers were in bloom yet, but their stalks were heavy with buds. Sallie pointed to the right, where two more mule deer were drinking. Their water source, Lin pointed out, was a kettle, a small body of water that fills a depression when blocks of glacial ice finally melt. We passed several ponds as we crossed the broad expanse.

Several times Lin pointed to a lone granite rock amid the green. These were

A circle of sunshine dapples an adult mule deer as it feeds in woods near the John Muir Trail through Little Yosemite Valley. Mule deer also browse in or near Yosemite meadows.

glacial erratics, boulders and rubble frozen long ago in a glacier's moving ice, then deposited when it melted. Some erratics composed of Cathedral Peak granodiorite made their way all the way to Bridalveil Meadow, on the valley floor. To the east I saw a needle-shaped spire—the top of Cathedral Peak. Here, Lin explained, is a perfect example of a trimline: the distinct line on a mountain that marks the upper limit reached by a former glacier. Cathedral's jagged spire, which rose above the ice, sits on granite shoulders smoothly rounded by the glacier's passing.

Ascending again, we climbed through remnants of snow pink with algae that smelled like fruit. I slowed. "It's called watermelon snow," advised Sallie, as she passed me, her pace constant and easy. Ahead, she pointed out dimples in a snowbank. "Sun cups," she explained, "depressions melted by the sun." The white granite crests of Cathedral Peak came into view east of the trail. We left the meadow far behind, and the trail grew steeper.

Snowfields clung to the peaks around us as we passed Upper Cathedral Lake, a deep-blue sheet rimmed by granite shelves and ledges. The trail headed upward across a marshy meadow, and I stopped to catch my breath. This was Cathedral Pass—the corridor Jim Snyder had told me Indians used for centuries to cross the Sierra. Behind us, in profile, Cathedral Peak revealed two towers. To the southwest, streaked with snow, the ragged ridges of Tressider Peak loomed. To the east rose the tall and barren ramparts of Echo Peaks. The walls were rough and chipped, some concave in spots—an example, Lin explained, of a process called exfoliation, or sheeting.

Granitic rock forms under great pressure miles beneath Earth's surface. It expands once it reaches the air. Joints—or cracks commonly found as sets of parallel fractures—are areas of weakness in the otherwise erosion-resistant rock. They admit water and air, which encourages weathering. As the rock weathers, space increases in the joints, and concentric plates, pieces, or slabs of granite exfoliate like onion layers. Joints usually follow topographic surfaces, so exfoliation usually parallels the topographic surface. Thus, sheeting is horizontal on a sheer rock face, convex on a dome, and concave on a valley floor.

Beyond the pass, we crossed another boggy meadow with wildflowers eager to bloom, then left it far behind and entered a world of gray interspersed with scraggly trees. All around us tumbled granite domes and peaks. To the east Lin sighted an arête, a sharp-crested ridge like a giant razor's edge sculpted by surrounding glaciers. The panoramas were incredible, but they were wasted on me as I grew slower and sicker from exertion and altitude. I was looking straight across to neighboring mountains, not up at them. No flowers here, just low sedge-like plants and thick mosses. No people either. Mac and Lin forged ahead to ensure that camp was set up by dark. Sallie, always unflappable, kept me company as I struggled steeply up, down, then up again. She pointed to a granite megalith called Columbia Finger. "Around that, then down," she encouraged me. And down it was, tight switchbacks through stands of pines to a high swale named Long Meadow. It was very long, more than a mile, but gently sloping, and we reached our campsite, Sunrise High Sierra Camp, at about 7:30.

We had hiked about seven miles; I was so done in I regretted every sedentary moment of my life since childhood. The next hour was a flurry of setting up

California's Yosemite Jewel

camp, pumping creek water through our filter to purify it for drinking, and cook-
ing dinner. Companionably tired, we ate, shared tea, and stowed every smelly
item from sunscreen to food into large bear-safe containers. Attracted to human
smells, bears are a continuing problem in Yosemite, so everything that might attract
them must be stored in lockers.

We snuggled into our sleeping bags and I dozed off immediately but wakened
to Mac's shouting: "Sallie, Sallie, what is it?" Irritably I sat up and looked outside
the tent. It was too dark to see anything, but I heard Sallie's quiet "Shoo, bear,
shoo." The animal was nosing at the tent flap until Sallie calmly sent it on its way.
As I said: She's totally unflappable.

IT WAS LIGHT WHEN I WOKE, ALMOST SEVEN,
and my three companions had been hard at work, getting water and starting breakfast. The sun was bright, and there was a fresh breeze. By the time we ate
breakfast, packed our gear, and set off, it was 9:30. We continued on the John
Muir Trail, past the path to Clouds Rest, and along a long and narrow knife-edged
ridge. At about 11:30 we started down a steep switchback trail, and along the way
we encountered two fellow hikers—the first people we had seen since leaving
Tuolumne Meadows the day before. The men told us they were hiking the Pacific Crest Trail, a 2,650-mile route from Mexico to Canada. Since leaving Mexico in
April, they had walked a thousand miles.

A shiny patch of glacial polish flashed on one side of the trail. I was accus-
tomed to the elevation now—nearly 10,000 feet—and was more easily noticing the
subtle differences in the rocks we passed. The bulk of the minerals in plutonic
rocks in the Sierra Nevada batholith are quartz, feldspar, biotite, and hornblende.
Lin showed me hexagonal flakes of biotite in some samples and dark rectangles
of hornblende in others.

We stopped for lunch at a little creek, and I gratefully soaked my feet in the
cold water. Snow plants, which feed on decayed matter, looked like red ice cream
cones. Lupines heavy with blue buds nodded near grayish-pink pussy paws. The
trail wound gradually down. The lupines on a ledge overlooking Little Yosemite

Particles of rock moving under thousands of tons of ice scraped the granite near Little Yosemite Valley, leaving shiny areas called glacial polish and lines showing the direction of the glacier's flow.

Valley were in full bloom amid tiny yellow flowers reminiscent of buttercups and tall Sierra wallflowers, their blossoms bright yellow balls.

At the trail's junction with the two-mile path leading to Half Dome, Lin and I dropped our packs and veered upward, toward the granite monolith that challenges thousands of rock climbers each year. We joined a motley group of hikers making the upward trek. Boy Scouts carrying duffels chugged manfully straight up; two women in sandals picked their way daintily through the dust; parents hefted toddlers; I trudged. Through stands of pines and over rock spurs, Lin and I worked our way to an area she called "the little bump"—the shoulder of Half Dome. Intrepid hikers picked their way farther, beyond steep piles of boulders and straight up the last 400 feet of the ascent, where they hauled themselves to the top using steel cables fastened to the sloping eastern shoulder. I could see those cables and human silhouettes on the flat top of the dome, but for me "the little bump" was far enough. We hiked back down to rejoin Sallie and Mac.

Backpackers ascend steep switchbacks above the Merced River on their way toward Nevada Fall.
Looming over the waterfall at 7,076 feet above sea level, Liberty Cap
(right, behind a glacial erratic) rises on the north side of the John Muir Trail.

California's Yosemite Jewel

The trail switchbacked down through stands of orange-barked incense cedars and pine trees. Sunrise Creek flashed to the left, and the sweet smell of western azaleas advertised the white blooms before they appeared beside the trail. People traffic increased. The private wilderness we had enjoyed the night before was nearly 13 miles away. By late afternoon we reached Little Yosemite Valley, one of the most popular campsites in the park. In contrast to our previous solitude, this was like a KOA campground on a holiday weekend. Beautiful, yes, on the banks of the Merced and among tall trees and deep duff—but chock-a-block with tents and people.

SATURDAY MORNING, EAGER TO AVOID THE packs of people that would be on the trail by midday, we decamped and set out early. We wound downward alongside the green and gurgling Merced River past cedars, pines, and nodding lupines. At a bench above Nevada Fall, we left the John Muir and took a steep spur trail along the north bank of the Merced. The path, a jumble of rocks and boulders, descended steeply to the base of the falls, and I leaned heavily on my walking stick for balance and to twist out of the way of day hikers. At the bottom, I scrambled over more boulders to look back at the Merced's roaring 596-foot drop.

We crossed a bridge to the south side of the river. Below us the rushing water bent back on itself then swept through a long, slanting ramp called the Silver Apron. Next the waters of the Merced collected in a deep green depression called the Emerald Pool, then sped toward 317-foot Vernal Fall. We stopped at about 10:30 for a snack at the bridge. Brazen squirrels eyed my crackers and string cheese, and I ate fast. By now water was precious, and I drank the last in my bottle. Far below, on the steep and narrow Mist Trail zigzagging up from the valley, a tour group clad in yellow slickers looked like baby ducklings. We took the Mist Trail down, and even with stone steps and a handrail, it was a tough walk. The route is very steep and bends sharply over huge stone blocks that are slippery with spray. Hordes of day hikers heading up from the valley made negotiating the turns difficult, especially since we were burdened with 40-pound packs. The waterfall peeped from breaks in the trees, and its thunder surrounded us. Mist cooled as the day's heat increased, and rainbows shimmered over the river. The scenery, as always, was spectacular, but the price was high. The traffic was constant; fat people, skinny people; Japanese, Europeans; kids, babies, grandparents; sandal-clad, sneakered, booted, and barefoot. The trail became a parade of pushing people.

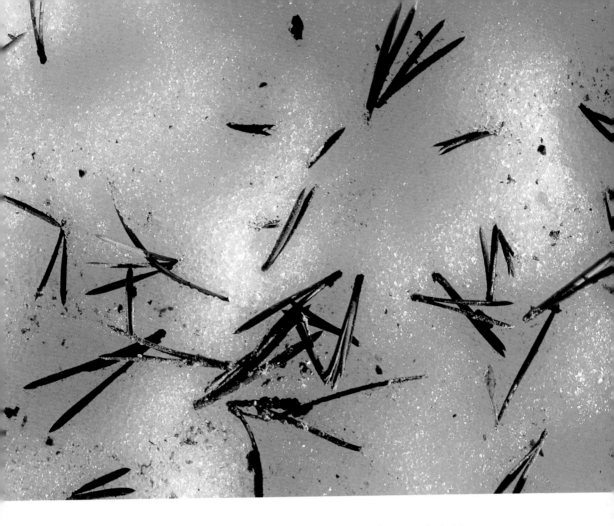

*In mid-June fallen pine needles scatter a snowbank near Cathedral Pass,
at an elevation of almost 10,000 feet. Snowfall in the High Sierra averages
more than a hundred inches a year. Drifts cling to shady slopes into late summer.*

Another bridge returned us to the north side of the Merced, and we walked beneath black oaks and pines. The trail sloped gently past enormous lichen-covered boulders, then grew rough and rocky as it approached the bus stop at the Happy Isles Nature Center. At about three, we hoisted our packs onto the shuttle for the short ride to Curry Village. We stowed our gear and sat on the terrace, rewarding our efforts with cold beer—the best I have ever tasted. Beyond the parking lot sparkled Yosemite Falls. Sun streaked Half Dome. I craned my neck to see Glacier Point looming above me and surveyed the chattering crowds around me. I knew firsthand what some of these people would never discover: The best hiking and the most enjoyment are to be found outside the valley, looking into it, where the people are fewer, the solitude is greater, and the scenery gives scope to the geologic wonders that make Yosemite so famous. ▪

TIOGA LAKE: *Gaylor Peak rises beyond the still waters of Tioga Lake, a mile from the edge of Tioga Pass, at 9,945 feet the highest automobile pass in the Sierra Nevada. A three-hour scramble up east of Tioga Lake leads to Yosemite's most accessible moraines, left by Dana Glacier.*

Conservationist John Muir extolled the Sierra Nevada as the *"most divinely beautiful of all the mountain-chains I have ever seen."* It is a young range of sharp, irregular granite peaks, roadless wilderness, towering sequoia and redwood groves, and wild rivers.

◆ TIOGA LAKE: Tioga Road enters the park at the pass that forms the eastern entrance. From it, hikers start daytrips and find trailheads leading to valley hikes and northbound backpack outings.

◆ BRIDALVEIL FALL: From a parking lot in the valley, tourists ascend a gently sloping paved trail a half-mile to the cascade; the round trip takes only about 20 minutes.

◆ MOUNT WHITNEY: The John Muir Trail starts on Whitney and winds 211 miles north along the crest of the Sierra, entering Yosemite at Donohue Pass and ending at Happy Isles in the valley.

◆ KINGS CANYON: The Pacific Crest National Scenic Trail, the nation's longest footpath, runs 2,650 miles from Mexico to the Canadian border, along the eastern portion of Kings Canyon.

BRIDALVEIL FALL: Morning sun softens the granodiorite ramparts
of Leaning Tower as Bridalveil free-falls 620 feet toward rising mist.
Like Bridalveil, Yosemite, Vernal, and Nevada Falls thunder
from hanging valleys left behind as glaciers melted.

California's Yosemite Jewel

MOUNT WHITNEY: Great range of granite mountains, the Sierra Nevada sweeps north and south from the high country of Mount Whitney, at 14,494 feet the highest peak in the United States outside Alaska.

California's Yosemite Jewel

KINGS CANYON: A packtrain picks its way down slippery slopes in Sequoia and Kings Canyon National Parks. Hikers traverse sharp peaks and snowfields at the mountain crests, alpine meadows and groves of giant sequoias at lower elevations.

California's High Sierra

CHAPTER FOUR

UTAH'S
DINOSAUR
COUNTRY

by Ron Fisher

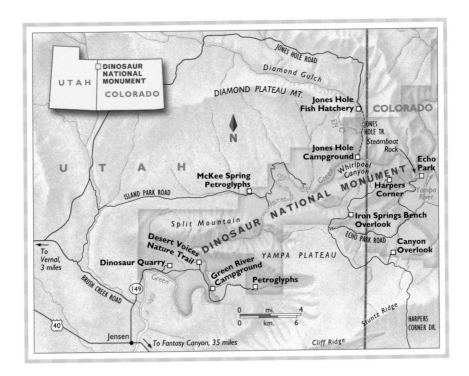

IKING AND GEOLOGY HAVE A LOT IN COMMON.

For one thing, they can both be slow and inefficient procedures. Airplanes and bulldozers would be quicker and more effective.

For another, in both hiking and geology, dogged determination pays off. Stick with it long enough and eventually you'll reach trail's end; unleash enough erosive forces and finally you'll carve a canyon. Take enough footsteps and you'll get to the top of the mountain; erode enough grains

Fossilized dinosaur bones 150 million years old survive in Utah.

PRECEDING PAGES: *Dusk highlights the rim of Split Mountain Canyon on the Green River in Dinosaur National Monument, which stretches across the Utah-Colorado border.*

of sand and the mountain will disappear. This was one of the realizations I came to as I recently visited America's dinosaur country in northeastern Utah and northwestern Colorado.

The landscape is dominated by layered rocks that come in many colors and configurations. They range, wrote Dr. Lehi Hintze, "from rocks formed more than two billion years ago to strata being laid down today. They are the book of stone that has recorded Utah's geologic history." And hiking in this region, where Utah meets Colorado—especially in the countryside around Dinosaur National Monument—is like leafing through the pages of that book.

D INOSAUR NATIONAL MONUMENT IS 210,844 acres—some 330 square miles—of rugged mountain and canyon country. Its gaudy geologic formations display a billion-year-old history of this part of the Earth. It's located where four physiographic provinces meet and overlap: the Wyoming Basin, the Rocky Mountains, the Colorado Plateau, and the Great Basin. Thus its more than 600 species of plants, 3,000 of insects, 21 of reptiles, 219 of birds, and 68 of mammals are representative of much of the West. Its natural communities include riparian, desert shrub, piñon pine and juniper forest, mountain shrub, and montane.

But it's the dinosaurs that led to the creation, in 1915, of Dinosaur National Monument. About 145 million years ago, this terrain—a low plain with rivers and streams crossing it—was just the right habitat for such creatures. During floods, the remains of dinosaurs swept down through overflowing rivers and landed, lodged, and settled on an enormous sandbar. While ages came and went, ancient seas piled thousands of feet of sediment atop them. The dinosaurs were preserved in the sand. Silica dissolved and sifted through the layers of remains. The strata hardened into sandstone, and the bones turned to fossils.

Dinosaur National Monument's geologic formations contain fossils spanning some 250 million years during the Paleozoic era's Permian period. The rising Rockies hoisted the region upward, and erosion gradually whittled it down, eventually exposing the bones. Western explorer John Wesley Powell reported "reptilian remains" here during his second trip down the Green River in 1871. Some 40 years later a paleontologist with the Carnegie Museum in Pittsburgh, Earl Douglass, began searching the area for dinosaur fossils. In 1909 he found "the weathered-out femur of a *Diplodocus*" and in 1908 a *Brontosaurus*, which he called "a beautiful sight." Andrew Carnegie contributed funds for digging, and Douglass moved his wife and baby out

A researcher painstakingly unearths dinosaur fossil bones. His workplace is a large wall of the Morrison formation displayed in the Dinosaur Quarry of Dinosaur National Monument.

Utah's Dinosaur Country

to be with him. Because Vernal, Utah, the nearest town, was too far to commute by horseback, they lived their first winter in a sort of tent on the site of the dig.

Douglass devoted his life to studying the fossils, and the site eventually produced a mass of dinosaur bones probably unequaled in the history of dinosaur collecting. Over 15 years, some 350 tons of bone representing ten species of dinosaurs were shipped off to museums. The bones were extracted from a rainbow-hued geologic formation named the Morrison, after a small town in Colorado near the site of the first discovery. Outcroppings of the Morrison formation appear in eight western states and have produced fossils from all four major late Jurassic dinosaur groups: plant-eating sauropods, stegosaurs, ornithopods, and flesh-eating theropods. The strata of this Mesozoic-era rock—red, maroon, purple, white, and green—consist of different types of clay, shale, and sandstone.

Douglass exposed a trench 600 feet long and 80 feet deep. After years of excavation, it was decided to leave the remaining fossils. In 1958 one sandstone wall was enclosed and became a museum display. Today it's the heart of the Dinosaur Quarry at Dinosaur National Monument. Created by President Woodrow Wilson in 1915, the original 80-acre monument grew in 1938 when the Green and Yampa river canyons were added to it. According to geologists, it offers the most thorough display of Earth history of any national park in the United States.

LOOKING FORWARD TO READING SOME OF Utah's book of stone, I spent some time in the northeastern corner of the state—and a little while in northwestern Colorado—following along with some geologists and walking a few of the many trails available to visitors. At Dinosaur Quarry, I took an open-air shuttle bus from the parking lot to the top of the ridge, where the museum sits. Inside, out of the hot sun, it was cool and quiet. More than 2,000 fossil bones are visible in the rock face of the sandstone wall opened up by Earl Douglass and on display here. A collection of the tools paleontologists use—chisels, picks, hammers—includes a bottle of Ecotrin, for creaky knees.

The bones on the quarry wall looked like leftovers from a children's game of

A silver ribbon, the Green River sunders the high country of Dinosaur National Monument.

PRECEDING PAGES: *Buff-colored, cross-bedded sandstone, formed from ancient sand dune deposits, rises alongside the Jones Creek Trail.*

A pictorial lizard shares the desert with the real thing: Fremont people created the pictograph above between A.D. 200 and 1250, pecking the image into desert varnish. The horned lizard below camouflages itself to blend into its desert habitat.

pick-up sticks, a confused jumble. Lots of small children were there the day I visited. A sign said "Touch But Don't Climb." Waiting for the shuttle bus back to the parking lot, a mother was trying to engender some enthusiasm among her overheated family: "You guys didn't seem very interested." The children looked a little abashed. One said: "I thought it was gonna be like a museum, where they've put 'em all back together."

There is such a museum, nearby in Vernal: the Utah Field House of Natural History State Park Museum. There, full-size dinosaur models roam a wooded garden, thrilling children and awing adults. The museum's curator, Dr. Sue Ann Bilbey, jaunty in her starched state park uniform, showed me through the back rooms where her newest—and proudest—find rested quietly on metal shelves: a nearly complete sauropod skeleton. The massive, blackened bones looked like they might have come from an especially large elephant. She showed me the end of the sauropod's tail, which had been nipped off, like a lizard's. "And see here?" she said. "Between these two vertebrae? Arthritis."

The Field House museum sits strategically in the Uinta Basin, a petroleum-rich intermountain valley south of the Uinta Mountains. Together the two features encompass nearly three billion years of Earth's history. Their fossils record more than 600 million years and include all forms of life, from algae to mammals. As they tromped through swamps and along muddy lakeshores, a surprising number of dinosaurs and other early creatures left footprints behind. Forty miles north of Vernal, at Flaming Gorge National Recreation Area, some tracks had recently been spotted. Several years of drought had lowered the water level of the reservoir, exposing them to view.

Sue Ann let me tag along with her and representatives from several U.S. and Utah agencies when they went to collect the newly exposed footprints. There were people from the Bureau of Reclamation (it was their dam); people from the U.S. Forest Service (it was their recreation area, and their boat took us from the Lucerne Valley Marina to the site); Utah state paleontologist James Kirkland; a graduate student and an intern; and representatives of the press from both Vernal and Salt Lake City.

Beneath a hot Utah sun, we stepped aboard a large pontoon boat and chugged across Flaming Gorge Reservoir, formed by a dam on the Green River. The name comes from a mention by Powell: He described "a flaring, brilliant red gorge . . . composed of bright vermilion rocks." We disembarked on a slope of sandstone slabs, just unstable enough that we had to watch our step.

When I saw the ancient creatures' tracks, I have to admit, I might have walked right past them. They look, to the uneducated eye, like natural deformations in

the rocks. The tracks of a giant sauropod looked like what would result if you dropped a bowling ball into mud.

Nearby are smaller, more delicate tracks of pterosaurs, birdlike flying and gliding creatures that are not properly dinosaurs but lived in the same era. Some were the size of crows, others as big as airplanes, with wingspans as wide as 33 feet. These pterosaur prints are three or four inches long, the size and shape of a small bird's silhouette—a sparrow, maybe. Alongside them are delicate marks made by the creatures' folded wings, which they used something like crutches as they walked awkwardly through the mud. Footprints from flying reptiles are rare, so these are prized.

The sun beats down, and the scientists murmur observations among themselves.

"This looks like it'll be quite a—quite a—find," says one.

"Nobody cares about the lower Morrison," says another. "Everybody wants the upper Jurassic."

"The sauropod tracks are significant because they've never been found this far north in Utah."

"This stuff would be about 150 million years old."

"There are several horizons here." (A horizon is a layer.)

A reporter observes, "But these rock layers are vertical."

Jim Kirkland, the Utah state paleontologist, responds: "A lot can happen to rocks in 150 million years."

By the time we break for lunch, a local radio reporter, hatless, has turned bright pink. Everyone finds a smooth rock to sit on and breaks out sandwiches, apples, cookies, and lots and lots of water. Listening to the conversations among them, I come to understand that paleontologists read tracks for information on all sorts of features of dinosaur life: family structure, social behavior, habitat, and physical characteristics such as size and body dynamics. The large sauropod whose tracks we were collecting was about 40 feet long and probably weighed 10 tons. To sit and stare at these tracks, imagining the landscape here 150 million years ago and trying to visualize extinct creatures trudging through the mud, gives everyone pause. These dinosaurs were like early hikers, I concluded, plodding along, leaving nothing but tracks. They were as slow and inevitable—and as ancient—as geologic forces. It's astonishing that those forces have preserved and protected these footprints down through the ages.

In hues that harmonize with the colors of the landscape's geology, lichens brighten a trailside boulder on the mile-and-a-half-long Desert Voices Nature Trail near the Dinosaur Quarry. In lichens, algal and fungal cells form a symbiotic relationship that allows survival on rock surfaces.

After lunch, sweating men work to excavate the footprints and send them on their way to the museum in Vernal for cleaning and display. They hoist each heavy slab of rock into a sling and muscle the slabs to the waiting boat. They set them carefully on life jackets to cushion them. Someone says, "Well, if the boat sinks, at least the rocks will be saved."

ANOTHER PAGE OF DINOSAUR COUNTRY'S book of stone can be read from the Harpers Corner Trail, within the monument. It leads to a spectacular view some 2,300 feet above Echo Park, where the Green and Yampa Rivers meet at Steamboat Rock and where once flat-lying layers of sediment were raised by the Laramide Uinta Mountain uplift.

Researcher Sallie Greenwood and I drove to the trailhead through gusting winds and a blizzard of Mormon crickets. The creatures—*Anabrus simplex* Haldeman, not true crickets—are each an inch or so long. Densities can grow gradually over several years, to a point where there may be a hundred Mormon crickets in one square yard. When populations reach such outbreak proportions, migration to foothills and rangeland begins. The creatures can't fly, so they migrate on foot. They walk and hop. They may cover up to a mile a day and travel as far as 50 miles in a single season. On this day, there were millions on the move, so thick, the roadway seems to shift sideways with their crazed and relentless movement.

Sallie and I would pass through Cretaceous, Jurassic, and Triassic rocks and pass by Paleozoic sandstone on our way up the Uinta anticline. We drove through patches of juniper and piñon pines, trees finely adapted to much of the arid West. We drove past charred snags from a 1972 fire. Ahead rose the derby-shaped Plug Hat Butte, whose variegated colors show its geological components. The hat's crown and brim are buff-colored Entrada and Glen Canyon sandstones; the red hatband is Carmel formation, all from the Jurassic period. In the distance was our old friend the Morrison formation; it gets its variety of colors from traces of iron, manganese, and other elements, and it's the formation where most dinosaur fossils are found.

From the Escalante Overlook, we peered down at the countryside crossed by Father Francisco Atanasio Dominquez and Father Silvestre Vélez de Escalante on their way from Santa Fe to California in 1776. When we crossed Wolf Creek fault, we were traveling on rocks older than dinosaurs. At Canyon Overlook we saw giant stairsteps in the Weber sandstone, where the Yampa and Red Rock faults have slipped downward. Ahead of us were the Uinta Mountains, a range of the Rockies. Around the time of the end of the age of dinosaurs, forces in the Earth's crust buckled once level rock layers upward into a broad arch, forming the core of these mountains.

At Iron Springs Bench Overlook, we saw some of everything that makes the area a geologist's textbook: great faults, rock layers now standing on end, winding canyons. At another overlook, we saw Steamboat Rock and, below us, Echo Park, a cliff-rimmed alcove of meadow at the junction of the Green and the Yampa. Here plans called for a dam in the '40s and '50s, but public opposition moved it a few miles north to Flaming Gorge. At the Harpers Corner trailhead parking lot, we waded through another tide of crickets, then followed the trail along a narrow forest ridge toward a rock promontory a mile away. A craggy piñon rose alongside the trail, home to insects and birds. Other trees showed evidence of gnawing porcupines. Far below us, on Pool Creek, a tributary of the Green, sat buildings of the Chew Ranch, home of a pioneer family who settled here around 1900.

Echo Park was named by Powell and his men, who rested here in 1869 and listened to their voices echoing off Steamboat Rock. He wrote, "All this volume of water, confined as it is, in a narrow channel and rushing with great velocity, is

Walking softly along Dinosaur's trails brings quiet rewards: a buzzing damselfly along the
Desert Voices Nature Trail, an alert mule deer just off the Jones Creek Trail. Explorer
John Wesley Powell named the creek and canyon for his photographer, Stephen Vandiver Jones.
The trail follows Jones Creek for four miles and ends at the Green River.

Hardy desert vegetation—a riparian, or streamside, community—lines the placid Green River. Some 200 million years ago, during the age of dinosaurs, the creatures roamed throughout the area, leaving bones and even footprints behind.

set eddying and spinning in whirlpools by projecting rocks and short curves, and the waters waltz their way through the canyon, making their own rippling, rushing, roaring music." A party of antlike rafters pulled ashore as we watched. The walls of Powell's Whirlpool Canyon are somber limestones and shales left by ancient seas. As the seas retreated, the wind piled up dunes that became the light-colored sandstone of the cliffs and domes along the Green River. That layer should be atop the Whirlpool Canyon rocks, but uplifting and faulting have bent and broken the layers and pushed some higher than others.

Sallie and I stood on the edge of one of the high blocks from which, long ago, the sandstone had been eroded away. A few Douglas firs, almost at the south end of their range here, rose alongside the trail on a north-facing slope. We searched the limestone rocks at our feet for the clamlike shells of brachiopods that have been found here, in rocks formed hundreds of millions of years ago, long before the dinosaurs.

WITH AN AFTERNOON YET BEFORE US, WE drove to another trailhead—the Desert Voices Nature Trail. We began in a parking lot, near the spot where Green River rafting parties put in and take out, across from Split Mountain. Tanned boatmen unpacked their rafts while their weary clients, in soggy sneakers, dragged themselves toward their waiting cars. Across the river, the warped strata of Split Mountain soared toward the sky. Here, the Green has split the mountain—hence the name—in a way that looks inexplicable. How can a river split a mountain? Powell theorized, "The river had the right of way." The mountain's strata once were level but got warped upward in an irregular dome shape. More sediments buried the dome, forming a nearly level plain on which the Green developed its course. As it began to cut downward, it struck the buried dome of Split Mountain. But, held in place by its banks, it couldn't change course and go around. Since then, erosion has exposed the mountain as a dome again, but now a dome split in half by the Green River.

The Desert Voices Trail, a manageable two-mile loop, has been interpreted by children—signposts along the trail bear their illustrations and writing. We pause in the hot sun to read them. One displays a picture of strata with the text: "I love layered rock. It was mud swirls that dried up and formed into rock. . . . It almost looks like a book almost like you could take and open it up."

Sam Perry, age 12, painted a peregrine falcon and wrote, "He sits there aware pondering what's there. Above him the sky swirls with excitement, its deep blue catches your eye. His side turned to a rainbow canyon, each layer a color. He sits there aware." A dozen or so of these charming signs line the trail.

From the parking lot, we backtracked to join the road that winds onward into the monument. From an overlook, we could see the Green River Campground, a lush patch of cottonwood trees shading the banks of the river. It looks very appealing and cool. Ahead and to the right, the hills were banded with shades of gray, red, purple, and brown—a sign that, once again, we had come upon the fossil-rich Morrison formation. A park service booklet says it looks like melting Neapolitan ice cream—and it does.

We passed Elephant Toes Butte, eroded from Glen Canyon sandstone, and some examples of Fremont rock art, where early inhabitants pecked away dark desert varnish to reveal the light-colored sandstone beneath. Signs warned you not to touch the pictures, but a family with boisterous little boys were all over the glyphs, caressing them.

At the absolute end of the road waits the sagging cabin of Josephine Bassett

Morris. After several marriages and children, Josie moved here in 1914 and lived alone for most of the next 50 years. She raised livestock—her corrals were fenced box canyons with nearly vertical walls that had eroded from Weber sandstone, a sand dune formation. She grew field crops, fruits, and vegetables. Her homesite is a cool and peaceful oasis with shading cottonwoods and Cub Creek trickling nearby. Still, I wonder—50 years alone here?

I found another lifelong resident of Utah's dinosaur country at the Field House museum in Vernal. Dee Hall, born in 1934, grew up in a sheepherding family a few miles north of Vernal. In fifth grade he visited the museum and impressed the director by correctly identifying the skull of a shrew. The director encouraged his interest and in fact became his mentor as Hall pursued a career in paleontology. Dee Hall worked as a preparator, preparing specimens at both Brigham Young and Harvard Universities, but now he's back in Vernal, retired and working as a volunteer in the museum's lab. He has devoted virtually his entire life to dinosaurs. "Sometimes they call me Dinosaur Dundee," he murmured with a smile.

Shy and courteous, soft-spoken and articulate, Dee took me for a tour of a couple of his favorite places around Vernal, including Fantasy Canyon, 40 miles south of town. It's a Bureau of Land Management site in the heart of the region's oil and gas production zone, so dirt roads built by the energy companies crisscross the region in a bewildering labyrinth. Petroleum here is found in the Weber formation, which normally lies between 4,000 and 6,000 feet below the surface. It's a stratum about a thousand feet thick.

Fantasy Canyon is ten acres of dull gray sandstone, eroded into fantastic shapes and forms. The rocks are from the Cenozoic's Eocene epoch, between 38 and 50 million years ago, deposited in Lake Uinta, which covered hundreds of square miles here. Its sediment of sand, silt, and clay formed into sandstone and shale, and over thousands of years the shale has weathered away, leaving mainly the more durable sandstone. The delicate shapes are constantly being reconfigured. Sign makers have succumbed to the apparently irresistible human impulse to give cute names to fanciful shapes, so Mickey Mouse, Jaws, a pirate, and various animals appear. A suspiciously recent Indian legend—it first appeared in print in the *Salt Lake Tribune* in 1972—claims the shapes in Fantasy Canyon are creatures of the underworld who climbed to the surface to conquer the Earth. They were frozen

---

The three-mile Sound of Silence Trail, near Dinosaur Quarry, invites visitors into a landscape of sandstone rocks and a sandy wash. Park authorities emphasize this is not a blazed trail but a route designed to help people learn to find their way in the desert—silence guaranteed.

Jones Hole Creek splashes in the Utah sunshine near its confluence with the Green River. The creek begins life as a spring four miles upstream at the Jones Hole Fish Hatchery.

in place when the world's greatest medicine man called upon the God of the North, the Great Lord of Ice and Snow, to immobilize the creatures with bitter cold, freezing them in the places where they stand today.

Erosion by weather and wind will eventually destroy them. Today Dee Hall and I stroll along the trail in blazing sunshine. The intricate forms and shapes look like they should be huge, but in fact they are tiny, so you feel like a giant walking among them. There are no dinosaur fossils here—the rocks are too young— but lots of fossilized reptiles and mammals. Dee walks with his head down and spots a fingernail-size fragment of an ancient animal. "Turtle," he says, pointing it out to me.

Fantasy Canyon is something of a team enterprise. The Bureau of Land Management oversees the area, the welcoming signboard was donated by Chevron, and the lettering was routed by shop students at Uintah High School in Vernal. The Utah Gas Company delineated the parking area with boulders, and Boy Scouts have placed formation markers, built a bench, laid a rock trail, installed a culvert, and made switchbacks.

Here in the West, where distances are great, the drive to and from a trail is often longer than the trail itself. We headed for home, talking as we drove past strata that swooped and swirled like cake batter gently folded and blended. Dee told me about his childhood fascination with porcupines. "I'd get hold of their tails, very carefully, then slip my hand underneath them." He once caught a young one, took it home in his sleeve, and raised it as pet. It lived outside, but it would come when he called and ride around in the basket of his bicycle. "Sometimes I still dream about porcupines," he admitted.

FOR MY LAST FORAY INTO DINOSAUR NATIONAL Monument, I chose the Jones Hole and Ely Creek hike, which some consider the prettiest of all. It follows along as Jones Hole Creek makes its way gently down from a spring, through a narrow canyon to its junction with the Green River. Powell named the canyon for his expedition's photographer, Stephen Vandiver Jones. The creek has eroded along a fault between the multicolored Lodore formation and buff-colored Weber sandstone.

The trail follows a path that probably has been used for thousands of years. It begins at Jones Hole Fish Hatchery, a place you don't so much arrive at as descend into. Down, down, down goes the highway, with signs alongside warning about the steep grade and urging motorists to shift down and to check their brakes. I set off down the trail fairly early in the morning, trying to beat the heat. Immediately the trail, the creek, and I are all in the cool shade of box elders. The trail is rocky but well trod. It is another day of extreme wind, and the trees overhead are swaying wildly. The creek is burbling at my elbow. Muskrat and mink have been spotted in it. It's four miles to the Green River and four miles back—so even though the trail changes only about 200 feet in elevation, I move steadily along.

Occasionally the trail rises slightly to cross open benches—level rock terraces that were once the creek's floodplain. These soils are drier, so bunch grasses, mountain mahogany, squawbush, and junipers grow here. From the benches I can see the canyon walls rising on either side of me. The left wall, which has been uplifted a thousand feet higher than the right, is formed of Madison limestone, a rock from the Paleozoic's Mississippian period, dating back some 330 to 360 million years ago. There are two formations to my right: the lower is of the Morgan formation, Pennsylvanian-age rock from 320 million years ago; above it is cream-colored Weber sandstone, 20 million years younger than the Morgan. The Morgan was deposited by an ancient ocean rich with life, so it contains coral, brachiopod, crinoid, and bryozoan fossils. The Weber was a sand dune. The Island Park fault fractured through here, weakening the rock and allowing Jones Hole Creek to do its job and cut the canyon.

After a mile and a half I come to a small wooden bridge across the creek and pause for a Rocky Road PowerBar and a rest. The creek makes a joyous gurgling and birds twitter overhead. A little farther down the trail, a signboard points me toward some pictographs made by Fremont Indians and a shelter used by them for some 7,000 years. Archaeologists named it Deluge Shelter after a flash flood in 1966 nearly swept them away.

Just after I read about this, a huge gust of wind in the treetops sounds exactly like a giant ocean wave, and I cast a wary eye on tiny Ely Creek. More dribble than creek, it comes in on my right, near a shady campsite where two men are putting up a tent. Occasionally a gust of wind is so strong, it picks up spray from the creek and showers me lightly—just enough to feel good. A small snake slithers out of my way, a creature I later identify as probably a yellow-bellied racer. Now and then I come upon fly fishermen. They are so focused on what they're doing, they barely nod as I pass. The creek and I cut ever more deeply into the Madison limestone.

Near the Green, I pass a rock outcrop that looks strikingly different: red sandstone of the older Lodore formation, Cambrian period—rock that's between 510 and 570 million years old. This ancient seafloor holds fossils of trilobites and brachiopods and marine worms. In that unimaginably distant time, such animals

Ely Creek waterfall tumbles toward a shady glen just off the Jones Hole Trail.
The Island Park fault crosses Jones Hole Canyon, giving the two walls a lopsided demeanor.

---

were only beginning to develop shells and skeletons that could fossilize. They predate the dinosaurs by hundreds of millions of years.

Once I reach the Green, I sit at a picnic table and share my lunch with some ants and a bee or two. Ubiquitous rafters slide noiselessly past on the river. I drink too much of my water and realize I will run short during the hike back to the fish hatchery. And during my walk back, I know, I will once again be leafing through dinosaur country's book of stone, one rocky page at a time. I'll make slow but steady progress, like a geologic force, but sooner or later I'll get where I'm going. ■

SOUTHWESTERN LANDSCAPES OFFER HIKING *opportunities with complex and dramatic geological forms: layered rocks dating from two billion years ago and strata accumulating even today.*

◆ GREEN RIVER: Hikers can exchange boots for boats to explore the mile-deep Desolation Canyon, upstream, on the Tavaputs Plateau of Utah. A 96-mile multiday hike parallels the river.

◆ GRAND CANYON NATIONAL PARK: America's world-famous 227-mile-long gorge, cut by the Colorado River, offers simple nature walks or strenuous treks to the bottom of the canyon.

◆ PARIA CANYON–VERMILION CLIFFS: This 112,500-acre wilderness straddles the Arizona-Utah border. Hikers explore slot canyons two feet wide and six hundred feet deep.

◆ ARCHES NATIONAL PARK: Hikers in Arches, near Moab, Utah, can explore more than 2,000 sandstone arches, as well as balanced rocks, fins, and pinnacles.

◆ LAKE POWELL: Colorado River tributaries offer a maze of side canyons for hiking: the San Juan River, biggest of all, and the Escalante River, which empties into Lake Powell.

GREEN RIVER: *Fossilized seashore ripples frozen in time lap the river bank. Some 7,000 rafters a year descend the river upstream from here, through Utah's Desolation Canyon, the tangled country where Butch and Sundance once hid out.*

GRAND CANYON NATIONAL PARK: A lone hiker contemplates tumbling waters and rising mists against limestone walls at Mooney Falls on Havasu Creek, along the three-day hike from Hualapai Hilltop to Beaver Falls.

PARIA CANYON–VERMILION CLIFFS: *Sandstone escarpment rises 3,000 feet, providing aeries for 20 species of raptors. Hikers sometimes spot deer and desert bighorn sheep here as well.*

ARCHES NATIONAL PARK: *Majestic South Window*
frames a pair of hikers and Turret Arch in the distance.
Some of Arches's trails lead walkers beneath
stone structures, offering dizzying upward views.

Utah's Dinosaur Country

157

LAKE POWELL: *Massive Jurassic sandstone walls dwarf a hiker. Nearby, hikers wade up to their ankles through the Escalante River.*

CHAPTER FIVE

MAINE'S ROCKY COAST

by Ron Fisher

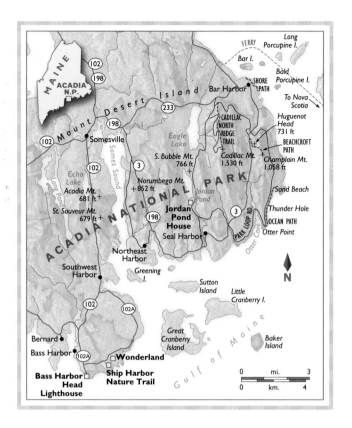

A map of Mount Desert Island showing Acadia National Park, with locations including Bar Harbor, Cadillac Mountain, Jordan Pond House, Northeast Harbor, Southwest Harbor, Somesville, Bernard, Bass Harbor, Wonderland, Ship Harbor Nature Trail, Bass Harbor Head Lighthouse, and the surrounding islands in the Gulf of Maine.

CADIA NATIONAL PARK IS AS HARD AS A ROCK—WHICH,
in fact, it is. A rock.

It was born 550 million years ago, in the
Cambrian period, when mud, sand, and volcanic ash accumulated at the rate of an inch a century on an ancient seafloor.
Powerful tectonic forces squeezed and heated the sediments
until they metamorphosed to schist.

One hundred fifty million years later, during the Silurian
and Devonian periods, tectonic plates converged, creating

Fog drifts through a spruce forest in Maine's Acadia National Park.

PRECEDING PAGES: *Sunset burnishes granite boulders on Acadia's shore. Massive ice sheets here
over thousands of years sculpted today's park.*

the granitic rock formations that eventually emerged on the surface of the Earth and became the core of Mount Desert Island. Mount Desert is a 112-square-mile isle just off the coast of Maine, lying some 35 miles southeast of Bangor. Mount Desert Island is home to much of Acadia National Park, whose dome-shaped highlands were once molten.

Now they are rock.

I was reminded of this one overcast morning as I set off to climb to the top of the park's namesake mountain—Acadia, 681 feet high, from the top of which good views of the surrounding park had been promised. Indeed, according to one guidebook, this walk "will culminate in one of the best views on the entire island."

Granite was underfoot as I began my climb. Granite steps lead from roadside upward into deep forest. A signboard—silhouetted hills and text handsomely mitered into a pillar—said ACADIA MT. .6 MI / 1.0 KM. I climbed steadily on a steep and rocky path. Drops of last night's rain dripped from the trees overhead. Almost immediately I came to large flat granite ledges, their centers dotted with cairns that marked the trail, their edges rimmed with low-lying blueberry bushes. I climbed through open spruce and cedar woods, crossed a brook and a fire road, then climbed up the steep clefts of a couple of massive ledges. The path turned to needle carpet among tall evergreens and switched back and forth over ledges and past pitch pines. A dove sitting on a boulder watched me pass but made no move. Blue blazes on the boulders showed the way. I came upon a family of four accompanied by a sort of wolfhound wearing a sweater and breathing as hard as I was. As I neared the top, I turned and caught my first views of two-mile-long Echo Lake, behind me, and the forested hills behind it.

A T THE TOP, ON THE NEARLY BARE LEDGE of Acadia's western and highest summit, I caught my breath and let the breeze cool me down. The view was grand. Tiny white sailboats dotted the surface of Somes Sound, and miniature motorboats chugged this way and that. Opposite me, to the east, the summit of Norumbega Mountain was hidden in hovering clouds.

A few feet away, across a small saddle, was the eastern summit of Acadia, and from there my view was 180 degrees. The white buildings of Somesville and Somes Harbor were in the distance. The sound looked like a big blue river as it stretched toward the Atlantic Ocean. Somes Sound was created fairly recently, in the grand scheme of things, by the geologic force responsible for Acadia's look and feel today: ice.

The Acadia Mountain Trail offers a view of Somes Sound,
a long, thin, glacier-carved, fjordlike body of water.

A day or two earlier I had stood on the eastern shore of the sound with Harold "Hal" Borns, Jr., professor of geological sciences and founder of the Institute for Quaternary and Climate Studies at the University of Maine at Orono. His special interest is in global climate change—tens of thousands or so years ago.

"When I went to college, I planned to study electrical engineering," he told me, "but in those days engineering majors were required to take a few courses in other fields—liberal arts or the humanities or some such. I signed up for a geology course

and never looked back." Hal has studied the geology of this part of the North American coast, among other places, for 50 years or so. He can recount the complex history of the region as if it all happened yesterday.

We stood on the shore of the sound as he sketched its convoluted record. "Maybe 20,000 to 18,000 years ago, the Acadian highland was a granite ridge running east to west. During the last glaciation, ice moved down from the north in huge sheets. At its maximum, it covered Acadia to a depth of maybe 8,000 feet. That's an ice sheet nearly two miles thick right here where we're standing! It extended about 370 miles out to the edge of the continental shelf. Then the climate warmed, and the ice margin retreated to the north, and cold ocean water flooded the valleys, about 14,000 years ago. The retreating ice left deep U-shaped valleys running north and south in Acadia and rounded and polished the mountaintops.

"Here, it gouged Somes Sound 130 feet deep below the present sea level. At the mouth of the sound, the glacier deposited a lot of gravel and rocky debris, and the water there is relatively shallow."

Hal and several colleagues have begun an effort to blaze an Ice Age Trail in this part of New England, patterned somewhat on Wisconsin's Ice Age National Scenic Trail. Maine's Ice Age Trail would lead visitors through an array of exposed glacial and emerged glacial marine deposits and landforms along Maine's east coast, explain their origin, and elucidate the Ice Age history of this part of North America. "The glacial features here are some of the best examples of their types to be found anywhere in the U.S.," Hal told me—especially compared to others on the East Coast. "They record the history of the last 14,000 to 13,000 years and reflect global climate changes in the North Atlantic region." The Ice Age Trail would begin atop Cadillac Mountain, here in the park—and so that's where Hal and I went next.

Though not a particularly high mountain, Cadillac is at 1,530 feet the centerpiece of Acadia National Park. Indeed, it is the highest point on the east coast of the United States. A very easy quarter-mile paved walk circles the summit and its large parking lot, and we strolled along it. On a clear day, according to a signboard, you can see Mount Katahdin, 120 miles away in central Maine. But today a thick fog swirled among the boulders. A little girl, remembering *The Lion King*, exclaimed, "Look, Daddy! Pride Rock!"

Reading a story in stones, Dr. Harold Borns of the University of Maine examines glacially quarried, polished, and striated bedrock south of Bar Harbor. Dr. Borns wrote the book on Acadia: He is co-author of The Geology of Mount Desert Island.

"If you had stood here 14,000 years ago," said Hal, "you'd have seen ice and ocean in every direction. The highest point of this mountain"—he softly tapped the top of a nearby ledge—"may have been the first part of the mountain to emerge from the ice."

It was an awesome scene to try to visualize.

"The enormous weight of all that ice would have depressed the land dramatically beneath it. We calculate that for every three feet of ice, the land would have subsided about a foot," Hal continued. "So, figure 8,000 or 9,000 feet of ice, the land would have been 3,000 feet lower than it is now. As the ice melted and retreated, the level of the sea rose, and the rebounding of the land would have occurred at the same time. But not exactly at the same speed, so we find evidence of beaches and sea caves up in the hills here, 250 feet above the present sea level. Acadia's mountains were islands."

Day's last light ignites a tall ship approaching Bar Harbor. From atop Cadillac Mountain, centerpiece of Acadia, hikers can see much of Mount Desert Island. Cadillac was built by tectonic and volcanic forces 420 to 425 millions of years ago, then shaped by glacial ice sheets.

Like many of the mountains of Acadia, Cadillac is asymmetrical: gentle and smooth on its north side, which faced into the ice flow, and steep and rugged on the south side. The distinctive shape is called whaleback—for its sperm whale profile—or roche moutonnée (sheeplike rock)—because it resembles a grazing sheep whose rear is higher than its head. As the melting ice sheets retreated down the gentle north sides of the mountains, they plucked large blocks from the south sides, accentuating the slopes.

Later, by myself, I hiked down the Cadillac North Ridge Trail to see that slope for myself. The town of Bar Harbor—a cluster of structures surrounded on three sides by water and backed by forest—sprawled below me, and little white boats dotted the blue harbor. I imagined myself facing into the oncoming ice. The trail descended fairly steeply, across flat slabs of granite marked with blue blazes and scrubby trees and bushes. Grasses grew in the clefts of the rocks. I was following an old road once traveled by horsedrawn buckboards. Now and then I brushed against the Summit Road that climbs to the top, then veered away again. Near the bottom I met a family on its way up. A little boy of four or five paused and looked around and said, "Dad, I think we're to the top." I knew he had a long way to go.

With Hal, I stopped by Jordan Pond House, famous for its popovers: The chef there bakes some 10,000 a day, and visitors gather at rustic wooden tables on the lawn to eat them with tea and strawberry jam.

"You can't quite tell it," said Hal, as we stood on a balcony overlooking the broad green lawn, "but we're atop Acadia's best known end moraine. The ice melted, and the meltwater carried sediments, as if they were on a conveyor belt, and released them at the front edge of the ice. This moraine forms a natural dam at the south end of Jordan Pond."

South of the pond, we stopped by a peacefully hushed cemetery near the town of Seal Harbor. Birds twittered melodically in trees nearby, and the cemetery, like all old graveyards, seemed to be quietly waiting. "Here was once sea level," said Hal. "Sand and gravel entering the ocean built up to the water's surface, forming flat-topped deposits called deltas. This one marks the approximate position of sea level when the delta was formed, about 14,000 years ago." Geologists study the fossils of sea creatures they find in glacial marine mud to date the formation of the deltas.

D RIVING AROUND ACADIA WAS ALL VERY well, but to get a close-up look at the park's geology, walking seemed more appropriate. There are approximately 110 miles of maintained, marked trails in Acadia. Most are at least 80 years old. None is more than a few miles long—the longest is 6.6 miles—so all can be done as day trips. Many climb rocky hills, but others edge the ocean or, like Jordan Pond, a lake. Only one or two are difficult enough to require climbing skills.

One of the easiest is the Shore Path, which begins at dockside in Bar Harbor and runs for half a mile along a harborside seawall. I walked it early one morning, passing a fine gazebo, its design copied from a famous one built in 1840 in Belfast, Maine. Guests at the elegant Bar Harbor Inn were sitting on balconies and lawn chairs, having their morning coffee and newspapers. Gulls hung about in a hopeful sort of way, looking for breakfast scraps. Magnificent homes line the shore here, and the masts of their docked sailboats swayed gently.

Fine-grained sedimentary rocks called the Bar Harbor formation appear on this shore, which inclines slightly downward into the water. Waves roll up and off

PRECEDING PAGES: *At 1,530 feet the highest point along the Atlantic Coast, Cadillac Mountain sometimes pierces the clouds with its peak. Subalpine plants such as cinquefoil take root here in the joints of rocks and among boulders. Wild blueberries and stunted, gnarled trees also survive.*

them. Some are slightly scarred with glacial scratches and grooves. The rocks of the Bar Harbor formation are the second oldest on the island, about 100 million years younger than the Ellsworth schist formed 550 million years ago. Offshore lie islands: spruce-topped Sheep Porcupine, Burnt Porcupine, Long Porcupine, and Bald Porcupine, all named for the slightly prickly appearance they share.

An even easier trail takes strollers 1.4 miles out and back to Bar Island along the gravel bar that gives Bar Harbor its name. Visitors can walk this trail only at low tide, so they learn to keep an eye on the tide tables. Children were exploring tide pools the day I strolled out there, and dog walkers were exercising their pets. Across the harbor a car ferry from Nova Scotia pulled in, turning around in its cumbersome way and backing into the dock. Bicyclists came and went, and a party of kayakers beached their craft, then loaded them onto a trailer and drove off.

ACADIA NATIONAL PARK BEGAN LIFE IN a fortuitous way, as a playground for the wealthy. From roughly 1880 until the Great Depression, the Rockefellers, Morgans, Vanderbilts, Fords, Carnegies, Pulitzers, and Astors built summer homes here. Like their brethren in Rhode Island, they coyly termed these homes their "cottages." Dreadful fires on Mount Desert Island in 1947 destroyed most of the cottages and brought the luxurious era to an end.

The park occupies about 50 percent of Mount Desert Island, the largest of the Atlantic rock islands. The island is 14 miles long and 8 miles wide. Eight of its mountains are higher than a thousand feet. It was named by Samuel de Champlain in 1604 for the treeless summits—l'Isle des Monts-déserts: the island of bare mountains.

John D. Rockefeller, Jr., donated nearly 11,000 acres of private estate to the reserve, which was designated a park in 1916. Today it's the sixth smallest national park in the country, but it has a comparatively high visitation rate, welcoming about three million visitors a year. It's home to some 50 species of mammals and more than 300 species of birds; it is also one of the sites of the attempted restoration of peregrine falcon populations into the wild. The park reintroduced 22 chicks

In late afternoon, a young zoologist investigates a tidal pool, discovering the miniature marine community alive there. Hikers reach the rocky edge of Otter Cove by way of the Ocean Trail, which follows the shoreline south of Bar Harbor.

in the park between 1984 and 1986, and each year since 1991, a pair has success-fully raised a family.

An era of active trail building had begun by the late 1800s. Village Improve-ment Societies in the communities sponsored much of the enterprise. Many trails were memorial paths, named by benefactors for people they wanted to honor. Early in the 20th century, trail builders began incorporating stone stairways and iron rung ladders into trails to traverse cliffs, slopes of rock debris, and other tricky areas. By 1915 there were more than 200 miles of trails on the island.

One of them, the Ocean Path, was built in the late 1890s and restored by the Civilian Conservation Corps in the 1930s. Today it runs for much of its four-mile round-trip length alongside a section of the park's loop road, a scenic twenty-mile circle that connects drivers to lakes, mountains, and seashore in the park. It began life as a buckboard road in the 1870s.

Along the Ocean Path, an ankle-deep gull scavenges in a shallow pool. Rugosa roses, left, blossom between pink granite boulders. Coves at nearby Sand Beach often display deposits of marine clay: flourlike particles abraded by glaciers, evidence that the coast was once submerged by the weight of the ice.

Weathered signs point the way at a junction on the Acadia Mountain Trail. A right turn here will take hikers to the top of St. Sauveur Mountain. Around the beginning of the 20th century, many Mount Desert Island communities began building hiking trails, and most are still in use today.

I set off one morning to hike the Ocean Path and parked at Sand Beach—a curious phenomenon for this area, where sand beaches are rare because the terrain is too young to have been formed by rock ground to sand. Here, shore currents have deposited sand and mixed into it the crushed shells of billions of whelks, periwinkles, crabs, sea urchins, and other marine creatures.

With the shrieks of swimming children resounding behind me—they may have been shrieking because of the nippy temperature of the ocean—I set off down the trail. Autos on the loop road zoomed by at my right elbow, and the ocean and its rocky beaches were on my left. Short, well-worn side paths took me over the rough granite ledges nearer the sea. Bayberry bushes and low pitch pines grew between the ledges.

After half a mile I came to Thunder Hole, a narrow chasm where waves boom when they rush in. It was quiet during my visit, but dozens of people were lined up, waiting expectantly, hoping to hear that boom. I met a woman hiking the other way with a ski pole in each hand. Offshore, black guillemots—a smallish seabird of the auk family—rode up and down on the swells: Up goes the swell, up go the guillemots; down comes the swell, down come the guillemots. I passed a beach composed of cobblestones the size of bowling balls, and in the distance I could see rock climbers at Otter Point, whose cliffs rise a hundred feet from the beach. There a mass of fine-grained sedimentary material is completely isolated in the granite: It was a huge fragment of the Bar Harbor formation that sank into the magma, recrystallized by the heat of the magma that enclosed it to a brittle quartzite. It evidently makes for a fine climbing surface, for it was a busy place on that sunny afternoon.

The Wonderland Trail is barely long enough—1.4 miles round trip—to qualify as a trail, but, as I found, it makes for a nice walk. It's near the very southern tip of the park so tends to be less crowded, and indeed I had it to myself on a breezy afternoon in mid-July. It follows a small peninsula that stretches into the sea, and a wide gravel path leads first into woods, then directly to a cobble beach facing Bennets Cove. A dove took flight with that curious whistling sound their wings make. A cluster of rugosa roses flourished just back from the beach. I sat for a while on a pink granite boulder, let the sea breeze keep the mosquitoes off, and watched lobster trap buoys in fluorescent colors bob in the swells. When the boulder began to feel too hard, I moved on.

Another peaceful trail—the Ship Harbor Nature Trail—is just a couple of miles farther along the coast, so I made it my next stop. According to an old legend, during the Revolutionary War an American ship, fleeing from the British, ran aground in the mud here—hence the name.

But people were here long before that. Encampments of American Indians have been found in Acadia dating back 6,000 years. European traders described Indians who lived by hunting, fishing, collecting shellfish, and gathering plants and berries. The Wabanaki Indians called Mount Desert Island Pemetic, the sloping land. According to historical notes, they wintered in the interior forests and spent their summers near the coast, probably here. But archaeological evidence suggests the opposite, that they summered inland and wintered on the coast, in order to avoid the harsher inland weather.

The Ship Harbor Nature Trail makes a figure eight as it wends its way from the parking lot to the beach. The well-marked trail is smooth except for occasional roots and rocks. Apple trees near the parking lot hint that this was once farmland, but alders are now crowding out the apples. A toppled, windblown tree farther along hinted at glaciers: They left behind bedrock and gravel, but shallow soils. Spruces, with spreading root systems, can grow in this thin soil but are easily blown over in high winds. I walked across huge slabs of pink granite and eyed the harbor channel.

At some point I took a wrong turn, evidently, for suddenly the numbers on the marker posts were running backwards. I slowed down to let a family of four get ahead of me. The place was busy with workers, young men and women—perhaps summer interns—rebuilding the trail, installing anti-erosion structures, and revegetating heavily used areas.

I caught up with the family, who turned out to be charming, despite their noisy children. They were visiting from Baltimore and preferred the quiet side of the park, away from the crowds and noise of Bar Harbor. I listened as the mother read the description of this area in the Park Service trail guide: "This channel leading into Ship Harbor dramatically displays the rise and fall of tides." Tides here empty and fill Ship Harbor twice a day. The water may be eight inches or eight feet deep.

At the beach, huge granite boulders offered fine seats. We watched a lobster boat working among the traps. The crew would maneuver the boat up to a buoy, pull the buoy and its attached trap aboard, empty it, toss it back, then move on to the next. "There's tonight's dinner," said the dad.

WITH A PARK RANGER NAMED CARRIE AND a handful of other tourists, I made a climb on another day along the Beachcroft Path to the top of Huguenot Head, elevation 731 feet. From Bar Harbor, Huguenot Head appears as a perfect dome.

The Park Service labels the walk "Blanket of Ice," and it's meant to acquaint visitors with the area's glacial past. "Ascend the steep side of an ice-carved valley to read the glacial record," the guidebook invites, and that's just what we did. We gained a hundred feet of elevation for every tenth of a mile.

A boulder beach in Acadia catches the dawn light. Endless surf has rounded and polished the stones. Otter Cliff, a popular spot with rock climbers, rises abruptly in the distance.

Weathered blocks of fine-grained granite, scarred by joints and fractures, front the beach at
Schoodic Point. A small outpost of the park exists here, on a namesake peninsula
six miles east of Mount Desert Island across Frenchman Bay.

We set off up one of the most distinctive features of many Acadia trails: granite steps. They were laid by the Bar Harbor Village Improvement Association in 1915, supervised by George B. Dorr, the so-called Father of Acadia. Between 1901 and his death at 90 in 1944, Dorr worked to acquire land for Acadia, using both his inherited fortune and his friends in high places. He hired local men to work on trails throughout the park, and the granite steps they laid here on Huguenot and elsewhere are really remarkable. They're not as evenly placed as stairs at home, naturally, but still, what a job it must have been. Heavy granite cubes and slabs have been smoothed and shaped and muscled into position on the steep side of the mountain as the trail switchbacks toward the peak.

CARRIE LED HER PUFFING FLOCK STEADILY upward, pausing now and then at roomy ledges to fill us in on the geology all around. At our second stop we could see, far below us, the parking lot at the north end of The Tarn, where we had parked—a narrow but deep pond left behind by the glacier. Already the cars looked very small. "The Beachcroft Path was named after the wealthy summer resident who financed its construction," said Carrie. "It consists of between 1,400 and 1,500 steps, all hewn by hand from local granite."

Carrie drilled us like schoolchildren.

"One of the major players in erosion would be . . . what?"

Water.

"Water! Right. And what kind of water?"

Ice.

"Ice! Correct. And here the ice took what form?"

Glaciers.

"Glaciers! Right. And we're talking about huge ice sheets. *Huge.* The Laurentian Ice Sheet moved right through here and covered the entire state to a depth of several thousand feet."

She told us how Cadillac Mountain came to be named. "In 1688 Antoine La Mothe de Cadillac, who liked to call himself a lord, arrived, having been given Mount Desert Island by Louis XIV. He only stayed here part of a summer, then moved on to . . . can anyone guess where?"

Detroit.

"Detroit! Right. He was put in charge of the post at Mackinac in 1694 and obtained a grant of land to what is now Detroit. The coat of arms you see today on a Cadillac's hood is the one the lord devised for himself. He wasn't really of royal blood."

At dawn, scraggly spruce lean to leeward on a beach in Acadia. Some 17,000 acres of Mount Desert Island's forests and many wealthy residents' homes were destroyed by a devastating fire in 1947. The Bass Harbor Lighthouse, right, rises 56 feet above mean high water at the southwestern point of the island. A covered walkway attaches the lighthouse to the keeper's house.

———————

During rests we could look west toward Dorr Mountain, east to the summit of Champlain Mountain, and north toward Frenchman Bay, where Bar Harbor lay. After an hour or so of more climbing, we stood in the breeze atop Huguenot and gazed north across the ocean. From this height, we now could see a large Norwegian cruise ship anchored off Bar Harbor.

Coming down Huguenot, three teenagers in our group couldn't resist eating all the blueberries growing in lush beds alongside the trail. Both their father and Carrie had to tell them to stop. "Leave some for the birds," said Carrie.

CARRIE HAD SHOWN US A BOWLING ball-size boulder near the top of Huguenot that was unlike the granite it lay upon. She called it an erratic: a piece of rock that had been transported by the ice from somewhere else. Acadia's most famous erratic sits high atop South Bubble Mountain, and I made it the destination for my last hike. Another sunny morning, another cool July breeze, another lacing of the boots and donning of the day pack containing water, candy bar, apple, guidebook, and camera.

I began by climbing through a quiet beech grove. Part of the trail had been carefully built of log cribs, to help control erosion. They were like stair steps and made for easy climbing. It was barely half a mile to the top, so I was soon there. The prize was football-shaped Bubble Rock, an 18-foot-long, 10-foot-high boulder perched precariously on a ledge overlooking the loop road. It looked as if the merest nudge would send it tumbling into the valley—but it's been here for centuries.

According to geologists, the boulder is Lucerne granite, a coarse-grained type not found in any of the Mount Desert Island ledges. It was deposited by a glacier, they believe, about 14,000 years ago, from at least 20 miles northwest of here.

I sat in the sun for a while, watching the cloud shadows and the people and the soaring gulls. I was looking nearly straight down on the tops of the forest below. It made me feel a little dizzy. I could again see the parking lot where I had left my car, and it looked as if Bubble Rock, if it suddenly came loose, would tumble down directly upon it. While I was thinking this, the alarm in one of the cars in the lot down below me let loose with that irritating security system *honk*— and I quietly hoped *that* would be the car the boulder hit.

Bubble Rock was my last stop in Acadia. It seems a fitting end, for like the park itself, Bubble Rock is solid, handsome, substantial, intriguing, full of information, and permanent.

A rock. ∎

A lush carpet of flora—including wild sarsaparilla and wood ferns in a carpet of club moss—clothes a coniferous forest bog in Acadia. Many common wildflowers grow in Acadia: Lily-of-the-valley, bunchberry, goldthread, bluebeard lily, and starflower are a few.

The long Appalachian Mountain chain that stretches north from Alabama to Maine grew from tectonic collisions perhaps 300 to 600 million years ago. Today mountains in the North retain their ruggedness, and trails that wind among them are rich with a patina of history.

◆ BAXTER STATE PARK: This park's 200,000-plus acres of land began with a 1930 gift from Maine's Governor Percival P. Baxter. Two hundred miles of trails blaze a landscape of rhyolite and granite shaped by glaciers.

◆ MOUNT KATAHDIN: Baxter State Park's crown jewel, at 5,267 feet Katahdin is Maine's highest mountain. Its height comes from the erosion-resistant granitic caprock at its peak that covers more easily weathered rock below.

◆ MOUNT WASHINGTON: Highest point in the Northeast at 6,288 feet, Mount Washington dominates the 1,200 miles of trails in New Hampshire and Maine's 800,000-acre White Mountain National Forest.

◆ PEMAQUID POINT LIGHTHOUSE: On the southern Maine coast near Damariscotta, hikers can discover an 1835 lighthouse standing guard atop swirling beds of metamorphic rock hundreds of millions of years old.

BAXTER STATE PARK: *Dawn vapors drift above a pond where canoes wait. Glaciers passed through here some 14,000 years ago, leaving behind moraines, erratics, U-shaped valleys, and kettle ponds like this one.*

MOUNT KATAHDIN: *Climbing from the north woods, Mount Katahdin, in Baxter State Park, marks the end of the 2,000-mile-long Appalachian Trail. Cirques, or bowls, on its flank were glacier-carved.*

Maine's Rocky Coast

MOUNT WASHINGTON: *Howling March winds and arctic conditions slow climbers in New Hampshire. Winds atop Washington reached 231 miles an hour in 1934, setting a record for natural land-surface wind speed that still stands.*

Maine's Rocky Coast

PERMAQUID POINT LIGHTHOUSE: *Wave-washed igneous and thin stripes of metamorphic rock look like taffy pulled by giants, according to a local writer. In fact, it is exposed plutonic rocks striped with seabed sediments, originally horizontal and now turned on edge.*

Maine's Rocky Coast

GEOLOGIC TIME SCALE FOR SITES IN THIS BOOK

This geologic time scale displays the correlation among known events and their eras at the sites described in this book. Eons are subdivided into eras, periods, and epochs, named for locations of characteristic rocks and fossils. The scale is in reverse chronology, with recent events at the top and events going back in time ranging down the scale. To put this scale in a larger time context, consider that the Earth was formed about 4.6 billion years ago, in the Archean eon.

EON	YRS. AGO	ERA	PERIOD	EPOCH	EVENTS AT SITES
PHANEROZOIC	10,000 to present	CENOZOIC	Quaternary	Holocene	**G:** Ice begins retreat from Icy Strait, ca 1750. **Y:** Last glaciers melt. A lake forms and lake sediments fill bottom of U-shaped valley.
	1.6M–10,000			Pleistocene	**G:** Continental ice sheet covers region. **Y:** Tioga glacier cuts U-shaped valley. **H:** Hawaii forms, ca 400,000 years ago. **G:** Wisconsin age ice covers most of region, 20,000 years ago. **A:** Ice cover reaches maximum depth of 8,000 ft. Sea level rises and falls.
	5M–1.6M				**Y:** Ice covers high country. Glaciers move downhill, quarrying bedrock.
	24M–5M	MESOZOIC	Tertiary	Pliocene	
	38M–24M			Miocene	**H:** Kaui formed by hot spot in Pacific plate.
	55M–38M			Oligocene	**Y:** Uplift of Sierra Nevada begins.
	66M–55M			Eocene	**Y:** Merced River cuts V-shaped canyon.
				Paleocene	**Y:** North American and ocean plates converge throughout. Mesozoic. Granitic plutons are emplaced. **Y:** Half Dome granodiorite crystallizes. **H:** Pacific hot spot begins to form islands.
	138M–66M		Cretaceous		**D:** Dinosaurs inhabit landscape of low plains and meandering rivers. **D:** Morrison formation deposited.
	205M–138M		Jurassic		
	240M–205M		Triassic		
	290M–240M	PALEOZOIC	Permian		**D:** Weber sandstone deposited.
	330M–290M		Pennsylvanian		
	360M–330M		Mississippian		
	410M–360M		Devonian		**G:** Limestone formed in tropical seas.
	435M–410M		Silurian		**A:** Sea sediment deposited in region.
	500M–435M		Ordovician		
	570M–500M		Cambrian		**Y:** Seas cover the region. Marine sediments accumulate to form rock of Sierra Nevada. **A:** Sea sediments accumulate in region, forming Ellsworth schist. **D:** Lodore formation deposited, including trilobites and brachiopods.

Key: G=Glacier Bay National Park, Alaska   **Y**=Yosemite National Park, California
H=Hawaii Volcanoes National Park   **A**=Arcadia National Park, Maine
D=Dinosaur National Monument, Utah & Colorado   M=million, B=billion

Further Reading

On America's geology
Harris, Ann G. *Geology of National Parks.* Kendall/Hunt Publishing, 1998.
McPhee, John. *Annals of the Former World.* Farrar Straus & Giroux, 2000.

On Hawaii
Chisholm, Craig. *Hawaiian Hiking Trails.* Fernglen Press, 1994.
Decker, Barbara and Robert. *Road Guides to Hawaii's National Parks.* Double Decker, 1997.

On Glacier Bay
Jettmar, Karen. *Alaska's Glacier Bay: A Traveler's Guide.* Alaska Northwest Books, 1997.
Molnia, Bruce. *Glaciers of Alaska.* Alaska Geographic Society, 2001.
Muir, John. *Travels in Alaska.* Modern Library, 2002.

On Yosemite
Huber, N. King. *The Geologic Story of Yosemite National Park.* Yosemite Association, 1989.
McPhee, John. *Assembling California.* Noonday, 1994.
Muir, John. *To Yosemite and Beyond: Writings from the Years 1863-1875.* University of Utah Press, 1999.

On Dinosaur Country
Hintze, Lehi F. *Geologic History of Utah.* Brigham Young Department of Geology, 1988.
Kelsey, Michael R. *Canyon Hiking Guide to the Colorado Plateau.* Kelsey Publishing, 1999.
Taylor, Andrew M. *Guide to the Geology of Colorado.* Lode Mining Co., 1999.

On Acadia
Chapman, Carleton A. *The Geology of Acadia National Park.* Chatham Press, 1980.
Gilman, Richard A., Carleton Chapman, Thomas V. Lowell, and Harold W. Borns, Jr. *The Geology of Mount Desert Island.* Maine Geological Survey, 1988.
St. Germain, Tom. *Acadia's Hiking Guide.* Parkman Publications, 2000.

About the Authors

Toni Eugene, a National Geographic staff member for more than 20 years, now lives in Charlotte, North Carolina, where she is a freelance editor and writer. She has contributed to more than 20 National Geographic books. For *Hiking America's Geology*, she returned to Alaska after 30 years, visited Hawaii and Yosemite for the first time, and came to rue—then change—her once sedentary lifestyle.

Ron Fisher first wrote about hiking for National Geographic in *The Appalachian Trail* in 1972. He has authored, contributed to, and edited books for adults and children. The author of National Geographic's *America A.D. 1000,* he retired from the Society's Book Division in 1994 and works as a freelance writer and editor from his home in Arlington, Virginia.

To Learn More

For further information on the National Parks in this book, including how to make your plans to visit and hike them, consult the U.S. National Parks web site: www.nps.gov. You may also want to explore www..aqd.nps.gov/grd/, the National Park Service's website on geology in the parks.

Acknowledgments

Special thanks to Kathy Edwards, Scot Gediman, Karen Hales, James P. Kauahikaua, Mardi Lane, Bruce Molnia, Judy and Greg Streveler.

Illustration Credits

1, Carr Clifton; 2-3, Gordon Wiltsie; 4, Raymond Gehman; 6-7, Stone/Getty Images; 8-9, Tom Bean/CORBIS; 10-11, Dennis Flaherty; 12-13, Stone/Getty Images; 14-15, Michael Melford; 16-17, François Gohier/Photo Researchers; 18, 20, Ralph Lee Hopkins; 22-23, 24, George H. H. Huey; 26, Ralph Lee Hopkins; 27, George H. H. Huey; 28, 29, Ralph Lee Hopkins; 30, 32, G. Brad Lewis/Seapics.com; 34-35, Douglas Peebles/CORBIS; 36, Doug Perrine/Innerspace Visions; 37, Tim Davis/Photo Researchers; 38, G. Brad Lewis/Seapics.com; 40, Kerrick James; 41, Masa Ushioda/Innerspace Visions; 43, G. Brad Lewis/Seapics.com; 44-45, 45, Ralph Lee Hopkins; 46-47, David Alan Harvey; 48-49, Ralph Lee Hopkins; 50-51, FPG International/Getty Images; 52-53, 54, Carr Clifton; 57, Macduff Everton/CORBIS; 58-59, David Muench; 60, Matthias Breiter/Minden Pictures; 62, David Muench; 64-65, Kerrick James; 66, 69 (upper), Tom Bean; 69 (lower), Carr Clifton; 70-71, 72, 74, 74-75, Tom Bean; 76-77, Kerrick James; 79, Tomas Mangelsen/Minden Pictures; 80-81, Tom Bean; 81, Carr Clifton; 82-85, Carr Clifton/Minden Pictures; 86-87, Marc Muench/David Muench Photography; 88-89, Phil Schermeister; 90, Robert Mackinlay/Peter Arnold, Inc.; 93, 94-95, 97, 98, Phil Schermeister; 100, Carr Clifton/Minden Pictures; 102-103, 105, 106, Phil Schermeister; 108-109, Stone/Getty Images; 111, 112, 112-113, 115, Phil Schermeister; 116-117, Marc Muench/David Muench Photography; 118-119, Phil Schermeister; 120-121, Galen Rowell/Peter Arnold, Inc.; 122-123, Galen Rowell/CORBIS; 124-125, Jim Wark/Peter Arnold, Inc.; 126, David Muench; 129, David Muench; 130-131, 132, Phil Schermeister; 134 (upper), Dennis Flaherty; 134 (lower), Scott T. Smith/CORBIS; 136-137, 139, 141 (upper), 141 (lower), 142, 145, 146, 149, Phil Schermeister; 150-151, Carr Clifton; 152-153, Stone/Getty Images; 154-155, Jack Dykinga; 156-157, National Geographic/Getty Images; 158-159, Carr Clifton; 160-161, Tim Fitzharris/Minden Pictures; 162, Carr Clifton/Minden Pictures; 165, 166-167, 169, 170, 172-173, 175, 176, 177, 178, Raymond Gehman; 181, Pat O'Hara; 182, Raymond Gehman; 184, Jack Dykinga; 185, Michael Melford; 187, Jack Dykinga; 188-189, Stone/Getty Images; 190-191, David Muench; 192-193, Phil Schermeister; 194-195, Michael Melford.

Index

Boldface indicates illustrations.

HIKING AMERICA'S GEOLOGY

By *Toni Eugene and Ron Fisher*

PUBLISHED BY THE NATIONAL GEOGRAPHIC SOCIETY
JOHN M. FAHEY, JR. *President and Chief Executive Officer*
GILBERT M. GROSVENOR *Chairman of the Board*
NINA D. HOFFMAN *Executive Vice President*

PREPARED BY THE BOOK DIVISION
KEVIN MULROY *Vice President and Editor-in-Chief*
CHARLES KOGOD *Illustrations Director*
MARIANNE R. KOSZORUS *Design Director*

STAFF FOR THIS BOOK
BARBARA BROWNELL *Executive Editor*
SUSAN TYLER HITCHCOCK *Project and Text Editor*
PEGGY ARCHAMBAULT *Art Director*
MARK R. GODFREY *Illustrations Editor*
CARL MEHLER *Director of Maps*
SALLIE GREENWOOD *Researcher*
NICHOLAS P. ROSENBACH *Map Researcher and Editor*
MATT CHWASTYK *Map Production*
JANET DUSTIN *Illustrations Specialist*
GARY COLBERT *Production Director*
LEWIS BASSFORD *Production Project Manager*
DALE-MARIE HERRING, *Contributing Editors*
ELIZABETH BOOZ
MELISSA FARRIS *Cover Design*
CINDY MIN *Design Production*
CONNIE D. BINDER *Indexer*

MANUFACTURING AND QUALITY CONTROL
CHRISTOPHER A. LIEDEL *Chief Financial Officer*
PHILLIP L. SCHLOSSER *Managing Director*
VINCENT P. RYAN *Manager*

Library of Congress Cataloging-in-Publication Data
Eugene, Toni.
    Hiking America's Geology / By Toni Eugene and Ron Fisher
            p. cm.
    Includes index.
    ISBN: 0-7922-6148-8 (regular)—ISBN 0-7922-6419-6 (deluxe)
    1. Hiking—United States—Guidebooks. 2. United States—Guidebooks. 3. Geology—United States. I. Fisher, Ronald M. II. Title.

    GV199.4 .E84 2003
    917.304—dc21   2002191238

One of the world's largest non-profit scientific and educational organizations, the NATIONAL GEOGRAPHIC Society was founded in 1888 "for the increase and diffusion of geographic knowledge." Fulfilling this mission, the Society educates and inspires millions every day through its magazines, books, television programs, videos, maps and atlases, research grants, the NATIONAL GEOGRAPHIC Bee, teacher workshops, and innovative classroom materials. The Society is supported through membership dues, charitable gifts, and income from the sale of its educational products. This support is vital to NATIONAL GEOGRAPHIC's mission to increase global understanding and promote conservation of our planet through exploration, research, and education.

For more information, please call 1-800-NGS LINE (647-5463) or write to the following address:

NATIONAL GEOGRAPHIC Society
1145 17th Street N.W.
Washington, D.C. 20036-4688
U.S.A.

Visit the Society's Web site at www.nationalgeographic.com.

Composition for this book by the NATIONAL GEOGRAPHIC Book Division. Printed and bound by R. R. Donnelly & Sons, Willard, Ohio. Color separations by Quad/Imaging, D.C. Dust jacket printed by the Miken Co., Cheektowaga, New York.